THE ADVENTURES OF Mas'KéG MIKE

Lynn Bday

Suite 300 - 990 Fort St
Victoria, BC, Canada, V8V 3K2
www.friesenpress.com

First Edition — 2015

The Author and publisher assumes no responsibility for mistakes, exclusion of all interpretation of subject matter in this book.

The events and recommendation in this book are not to be taken as professional medical or any recognized therapeutic direction. All readers are advised to seek the services of competent professionals that have experience in the treatment and rehabilitation of an acquired brain injury. Thank you for reading it.

ISBN
978-1-4602-3842-4 (Hardcover)
978-1-4602-3843-1 (Paperback)
978-1-4602-3844-8 (eBook)

1. Biography & Autobiography

Distributed to the trade by The Ingram Book Company

Contents

Mâs'kég Mike - the man who "wore" with many hats

THE ADVENTURES OF MÂS'KÉG MIKE

The Free-Spirited Heart in the North

A MEMOIR

BY MICHAEL OUELLETTE

My stepfather Ernest Coucroche. 2004

DEDICATION

For COOKIE

A man of great wisdom and patriotism
For all Canadian Veterans – Past and Present
With special mention of our Aboriginal Veterans
This humble piece of work is in honour of you.
Your heroic deeds will never be forgotten.

"God and the soldier we alike adore, in times of danger not before!
The danger past and all conflict righted, God
is forgotten the soldier is slighted!"

— *Rudyard Kipling* —

and
For my family
To whom I will forever be grateful.

ACKNOWLEDGMENTS

My warmest appreciation to a veteran broadcaster and researcher, Mr. David Miller, who, for over 25 years, continue to increase public awareness on the preservation and appreciation of our cultural heritage in Yellowknife, Northwest Territories. With his permission to mention his real name in my book and talk about the interview he did with me live on CBC Radio in 1992 is a great honor. Later all our exploration crew was in a one-hour documentary at the same site.

I will always be grateful to Junie Prunie for creating a mental picture of my book, pushing me outside my comfort zone and believing in my ability to write a book. With permission from her son and closest friends, I have one of Junie's poems as a tribute to her – a writer and an artist in her own unique way. She never held back from giving advices whether during a visit to her apartment or over dinner and a cup of tea.

Ron Cudney, one of my closest buddies, inspired me to write from scratch and is the author of "Skinny" and "As A Matter of Fact". I've known him as Ace since the days up in the 'Knife and the Yukon. Ron has helped me with ideas and suggestions via email, phone chats and during visits since 1982. One thing we have in common - we are lucky husbands of lovely ladies from the Philippines.

Cathy Chiasson, an avid admirer of the Northern Lights from Yellowknife, Northwest Territories, captured the amazing beauty of the Northern Lights as featured on the front cover of my book.

My deep gratitude to the North Slave Métis Alliance for supporting my Business course at the Academy of Learning in Yellowknife, Northwest Territories, where my typing and computer skills were honed and made the tackling of my book writing quicker and easier.

Membership with the L'Association Franco-Culturelle de Yellowknife reminded me of my Aboriginal / French roots through arts and culture. In 1993 one article entitled "Working and living off the land in today's society" that was published in their newspaper, L'Aquilon, that featured a day of my life in the taxi business and working in the mines as Mâs'kég Mike was inspiring.

Without the help and support of the medical professionals at the Stanton Territorial Hospital in Yellowknife, Northwest Territories and those who tended to me in North Bay, Ontario, especially my family doctor and my neurologist, I would not be back on my feet again.

The Ontario March of Dimes' On with Life program (O.W.L.) in North Bay, Ontario continues to support people like me through group sessions and other activities like swimming and baking, making life more fun and manageable. Furthermore, the Taoist Tai Chi Society, the Brain Injury Association of North Bay and Area (BIANBA) and the North Bay Heart and Stroke Foundation were also helpful during my struggle towards recovery and continue to do so. The Heart and Stroke Foundation gave me the opportunity to ride the Big Bike three years in a row now since my first time in 2012 with the "Rebel Riders" Team.

The Knights of Columbus Council 12030 (Holy Name of Jesus Parish) of which I am a member helped me to ease back into community involvement by encouraging active participation in local events such as volunteering at the blood clinic and soup kitchens, and help prepare meals at our church to raise funds for the Order and its recipients, one of which are the Boy Scouts and Girl Scouts of Ontario.

The efficient staff of the North Bay Public Library contributed greatly during my research on our lineage and the equally efficient and friendly staff of the Harris Learning Library at the Canadore College in North Bay, Ontario shared their technical expertise when I was faced with technology monsters in their highly advanced computers that were not familiar to me.

My genealogy research started with the invaluable resource, a 12-generation custom-made genealogy book of both sides of my family, provided by the La Société Historique de Saint Boniface, Manitoba.

The very talented Mike Gauthier continues to be very generous in sharing his information and resources regarding our lineage and matters involving the Algonquin First Nation. In our research, it turned out that Mike and his wife, Debbie, are both my relations! I helped Mike build a birch bark canoe which went on display during the commemoration of the 100th anniversary of Anahareo's life in June 2006 and later at the Capitol Theatre in North Bay, Ontario. Many of Mike's handiwork are currently on display at the Mattawa Museum in Mattawa, Ontario.

I am thankful to the staff of FriesenPress, most specially Brelan Boyce, my account manager, whose patience, invaluable suggestions and expertise made this book a dream come true for me.

My special thanks to my brown-eyed girl and sweetheart of mine, Iris Margaret, who stood by me through thick and thin during the ups and downs of our roller-coaster life.

To my family and close friends, with you in mind, here's hoping you will find enjoyment in reading this book as I take you down memory lane.

INTRODUCTION

"There is no greater agony than bearing
an untold story within you."
– Maya Angelou

I am a survivor of a near death experience in my early fifties. On Groundhog Day 2011, I fought back from a traumatic brain injury to pen this witty memoir with the help of a few.

I had long wanted to write my life story but with no background in writing, it kept falling through the cracks. Until one day, a very dear family friend, the late June Ursula Kuehl, who I fondly called "Junie Prunie", encouraged me to put them fun stories into writing. She told me one day that my stories always amused her and sometimes almost made her pee her pants. There was the little nudge I needed that sent me digging for old photographs and to start weaving stories and ideas together to start my book. With a few pictures gathered from family and friends, I started scribbling and then later putting into good use the typing and computer skills I had started to learn in 1999 at the Academy of Learning in Yellowknife, Northwest Territories. "I started my book," I e-mailed Junie one day. She had just started to learn the use of a computer herself. "I've known you since you were nine years old keep on writing," she replied. "I know there are lots of stories in you waiting to be told." I have other close friends who had written books and I have personally met a few other good writers which boosted my confidence that I could definitely give writing a try.

Life is uncertain, we all know that. Here today, gone tomorrow. So before my brain forgets and the memories vanish like they never happened, I'll tell you a few short stories of where I've been. As Frank P. Thomas writes, "Your life is important. It is as unique as your fingerprints. It is a precious piece of time that should not be forgotten. There has only been one life lived like yours in this lifetime."

Visiting or having relatives visit us at home with a good game of cards was one of the favourite pastimes in the country, aside from playing baseball and soccer for the kids with cousins. Fishing and hunting when time allowed was another family summer and fall favourite but done so safely. The lazy days in the farm ended with family and friends gathered for tea or a cold one, and for sure a story-telling marathon. Over the years hearing so many of them, I enjoyed the stories as some of them where worth passing on whether I was doing the talking or at the receiving end of it. Many stories I've heard from relatives or strangers sometimes struck a chord in me and left imprints that later served as lessons of life growing up. This book is for those many people and places that shaped me in to who and where I am today. So here I would like to eagerly share many of the recollections I've gathered so far.

In your hand lies what I would consider one of my greatest accomplishments and definitely at the top of my bucket list. I'm a firm believer that with God all things are possible. My near-death experience was the turning point where I realized how vulnerable I am, and that life can be easily snatched from you when you least expect it. Life is too short to be selfish and to live in melancholy. If, with this book, I can make the world a little happier and inspired, page after page, why not? I invite you to trek along with me and let me take you to these places that provided the kaleidoscope of my book. NOTE: Some stories may have a familiar ring to them, or maybe not, but one or two may be about you! Names, places, and events in this book are factual, however, some character names were altered to protect their true identities.

These days, I'm chiefly living in the moment with no regrets about travelling the zigzag trails from coast to coast. I certainly did a little True Northern howling while on the burning Ring of Fire with a little panning for gold and diamonds. In those days, I had a humorously erratic moral arrow held and compassed in the stock and shares market of today.

~Mike~

My wife and I rode with the Rebel Riders Team during the Big Bike Ride in support of the Heart & Stroke Foundation with the Mayor in North Bay, ON

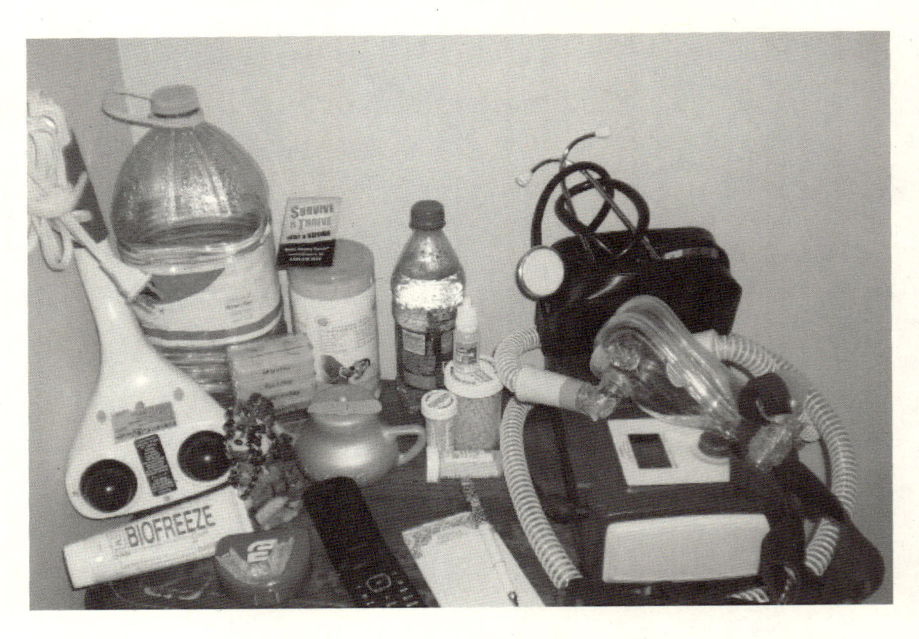

Each item on the table contributes to the journey of my recovery

CHAPTER 1
ACQUIRED BRAIN INJURY

My Ordeal With Acquired Brain Injury

Lac de Gras, Northwest Territories, 2011

It was a freezing -30°C outside at the isolated mine site workplace. Having been away from the mining profession for a few years for one reason or the other, now back for a better buck, I suddenly felt nostalgic seeing a familiar scene of a typical mine site like the many others I worked at before.

Having worked at the Joint Task Force North headquarters as a security officer with the Corp of Commissionaires up in Yellowknife in the Northwest Territories, dealing with people of all ages. Now working at this mine site, as it is usual to see people from different nationalities and backgrounds, my experience would make this job easier, or so I thought. Although all that walking I did while issuing parking tickets sure paid off as I started my twelve-hour shift at this mine site.

This was my second two-week stretch as a trainee at the crusher plant where diamond-bearing ore is crushed to separate the diamonds and other heavy minerals from the waste rock. All of a sudden, in the coffee room while on a break, a jerk coworker made a very discriminatory remark to my Aboriginal long-time friends. They were First Nations from of the Northern Arboreal Arctic regions of Canada and now were full time workers of this mine trying to make a decent wage just like the rest of us. In their defense, I stepped in and verbally reminded the jerk that name calling would not meet his bossy quota request. Getting ready to head back to work, he just smiled and said, "Come on, you lazy Indians."

I have been a hardworking, Yellowknife employee in the public service for over twenty-five years myself and a registered member of the Northern Metis Alliance of Yellowknife, NT. I currently hold a status with the Mattawa -North Bay Algonquin First Nation in Ontario, Canada. I was born and raised a Frenchman of Ontario and now married to a lovely Filipina lady. I do not go for racism nonsense at all. I reminded that irritating, contemptible racist that one cold evening the winter before while he was at the Yellowknife Inn he had been a taxi passenger of mine. The same fellow who was outside staggering and wobbling so much that I had jumped out of my taxi to assist him as part of our safe public service to all passengers. Should there be any doubt, we were to request dispatch to call the Royal Canadian Mounted Police (RCMP) immediately which I almost did. Drunken pedestrians do get picked up right away at 40 below zero.

As a Commissionaire in Yellowknife, NT. 2007-2009

The Joint Task Force North (JTFN) HQ. Yellowknife, NT

I asked him if he remembered the wretched moment that particular evening. As a registered first- aider at our taxi stand office, safety always comes first. I assisted this fellow to the passenger side of my cab, asking his buddies on the hotel stairs if he needed to go to the hospital.

"Oh no he'll be alright". "He's had had a few drinks. Just take him to the address he has in his coat pocket with twenty dollars," they said. "It's not far away. Keep the change," I was told.

I called dispatch to let them know this customer was said to be inebriated. He sure reeked of something else! Having driven taxi for many years my nasal scent is of better than average.

So I drove him to the address which was at an Aboriginal family's household I knew very well. With his arm over my shoulder, I walked him over, knocked on the door and then carried him all the way to the couch as he was practically unconscious.

Dispatch had sent me over to yet another call back at the Yellowknife Inn but the time I arrived competitors was helping another wobbler, a well-known local, into their taxi. Rather than get into a scrap over something as silly as this avoidable dispute, I went to speak to my supervisor and told him my side of the story. I showed him my Algonquin card and explained that what the other guy did was unacceptable and definitely not an appropriate attitude of a team player. As a sanction for causing a commotion, both the fellow with the snide, racist remark and I were both put into fire guard duty as part of our safety protocol. This new assignment was a physically demanding stair-climbing task for a twelve-hour work day.

The crusher plant was partially under repair and the fire alarms coincidentally were said to be out of order. So rather than a fist to cuffs for the outburst of a preliminary topic of a work week, it was better just to ease off. We were dressed accordingly for an extremely dusty and noisy environment while following all the work detail which was usually detected by a sound and a light warning system but now we had to do it by eye. We were watching for smoke, sparks, broken or loose parts while having a sharp eye for flying rocks.. Top of the list was an overall safety check for a tear or odd displacement of belts from loose rocks that might have been stuck underneath as they got crushed into smaller particles from one belt layer to the next, or anything that might be a fire hazard.

With the Step Back 5x5 workplace safety policy in mind, a note pad handy, and a two-way radio within reach in case of emergency or if fire hazards were detected, I was prepared. We had to do a morning check on the two-way radio by pressing the button to make sure it works and waiting for a reply to confirm my location. Otherwise I had to walk back to the office as need be.

Here now on our last day of the two-week stretch, I did get a radio call for an afternoon in-class training that was about to take place within the hour. "Roger, copy that", I said in response. I headed outside for the van that took me to class at another building. That sure will be a nice needed break, I thought. I headed down to the closest washroom. I took of my mask and ear muffs to wash off the dust. It seemed I was alone in this good sized washroom as I cleaned the thick dust off my face. Then as I stood by urinal I heard something that sounded like a spray. Suddenly, I smelled some sort of a bizarre, sweet odour of a strong substance quickly filling the air. I started coughing and felt dizzy and oddly sick as if ready to puke. Slowly turning around, I got a view of a pulled in set of boots at a closed door on one of the far-end stalls, I hurriedly grabbed my gear to leave the washroom headed outdoors and banged the stall door on my way out to that much needed fresh air. Suddenly a loss of balance caused me to fall on the ground and my lights went out. I was told later, in a report by the Workplace Safety Insurance Board (WSIB) that I was about fifty feet away from my supervisor's office to inform him of this odd smell in the washroom. I do not recall anything that happened after that until I woke up three days later in the Intensive Care Unit (ICU) of Stanton Territorial Hospital in Yellowknife.

Yellowknife, Northwest Territories

It was the day before Groundhog Day in 2011 that my life made a 360° turn. I wasn't prepared for it but here I was lying on a hospital bed with no idea of what the heck was going on.

Apparently, the twenty-eight days spent in the hospital had made me two days short of the thirty-day requirement to qualify for Long Term Disability. I was told before I was discharged that I would likely have to head back to work sometime this year. I had the use of a walker and ski poles for three months due to balance issues while taking a handful of medication each day I was not in any hurry that's for sure. With had the help of long-time friends for transportation during the bitterly cold winter months. I write this chapter with commitment, passing on my story of a near-death experience in the workplace in Canada for it not to ever happen to others in the future. The twenty-five years of working in the mines in and around the community of Yellowknife will speak for itself.

As I slowly recovered through physiotherapy and occupational therapy and the constant support of the medical staff I still suffered from short-term memory loss and blank moments and other complicated issues. The daily dose of medication slowed me down in all aspects of my life. On my first month I was on six pills of anti-convulsions but was increased to twelve a day. This gave me overwhelming side effects such as headaches that felt like an icepick poked through my right temple, severe chest pains short attention span, easily distracted, very poor balance, inability to tolerate loud noise or music, and shortness of breath – all these made me fearful. My body was pumped up with lots of medication that I did not particularly want to take as it was very hard on my system.

Staff from the mining office downtown came with forms to be filled out by the attending doctors as part of the workplace benefits. My loving wife and I went through all the written reports of what had happened to me. She acted as my power of attorney to sign forms on my behalf. Her patience to help me do so was well appreciated. Never in my life did she have to deal with such amount of paperwork.

In June of 2011 my wife and I decided to move back to Ontario after I weaned myself off of all these medication in a period of three weeks. I felt this was a healthier choice coupled with prayer to God for guidance.

With one letter I recounted all of these incidents accompanied with hair samples to the Workplace Safety and Insurance Board (WSIB), the Royal Canadian Mounted Police (RCMP) in Yellowknife, the Ministry of Development and Mines, and other agencies that I hoped would assist in putting the puzzle pieces together. Should a chemical exposure have caused my near-death experience, my work clothes and DNA samples would get checked. A WSIB on-site report claimed that viral encephalitis was the only culprit for all these.

With plans to have a child I completed three medical exams in less than a year prior to this incident as a job prerequisite and to show that my wife and I were healthy and truly ready to start planning for a new addition to our family. But all our hopes and efforts had went down the drain after this workplace incident. I've defended myself in small claims court more than once and won every one of them against liars through the power of a pen and honesty including parking tickets I issued that were contested in court. I am a firm believer that honesty will set you free.

As part of the Emergency Action Plan protocol, the medical officer at the mine site tried to call my wife through her cellphone but since cellphones were not allowed in her workplace, she did not get to speak with the doctor until after five o'clock. However they did reach my sister who lives in Ontario as she was the other contact person in case of emergency.

By 2012 all the information I gathered so far had been submitted to our legal system, government agencies, and Canadian Blood Services. I obtained a written permission from the Canadian Blood Service's main office to volunteer at one of their blood clinics through the Knights of Columbus. As a proud blood donor in Canada and in the United States before the incident only proves that I was never infected with viral encephalitis or had any medical condition that may prohibit me from ever donating blood again.

Part of my boot camp training with the Canadian Armed Forces in Cornwallis Nova Scotia was developing my skill as chef cook. We were put to test blindfolded and asked to identify herbs and spices by smelling. "Safety first" is key to survival and this entailed wearing proper safety equipment such as properly wearing a gas mask when handling poisonous toxic chemicals. I can say that during my twelve weeks at boot camp and even at the farm back home I had been exposed to various substances and was shown how to properly protect myself. With what I have learned and with common sense put together, I certainly knew the definition of smell better than the average person. I have been commended for the many safety hours in the workplace that I've done and I can honestly claim I always made sure to stay safe in each workplace I was at. I once owned a propane powered vehicle and had worked with many other gasses of highly volatile nature that are dangerous and could jeopardize the lives of workers. But despite my extensive exposure to these different chemicals never did I encounter such a strong smell like the one that hit me in the washroom that fateful day.

During the span of over ten years in the isolated mining regions I got paid fifty dollars extra per day as a level two first aider ready with notes to contact by radio for a flight out in cases of serious injuries. My personal support worker training in 2000 sure helped make a difference to save lives of coworkers. With over 20,000 in safety hours I received many cash prizes and bonus pays, one of them by Midwest Drilling of Manitoba. Even stool and urine samples were required as part of many jobs I applied for and I proudly claim that I passed each one of them with flying colors. This procedure also proves one's health.

Mishandling of oxygen such as administering to a person who does not require it may cause an embolism in the arteries to the brain and immediate loss of consciousness may occur, a convulsion, a mild to severe stroke, **even death**. These facts our taught to us from the start. The opportunity to help save lives of co-workers, including one in particular incident where one of the men almost lost his lower extremities due to serious frostbite was always there. If not for a quick remedy of gradually warming him up while trying to keep his blood circulation going to his frozen limbs, he would be in a wheelchair today. Being the site contact as first-aider of the isolated region, I was trained to have the radio close at hand so that during emergencies such as this one I was able to coordinate with the medical team in town for air and land ambulance to the hospital. The Workplace Hazardous Material Information System (WHMIS) controlled product regulation is part of workplace safety at all times. At the mines Jet B fuels were used as well as other chemicals used in the processing of ore are classified as hazardous. But proper handling is key to prevent accidents.

I have flown many hours in helping helicopters hook up loads, safely bringing fuel by the tank load. Once at camp, a co-worker dealing with a bad case of hangover, filled one of our diesel furnaces and lit it with Jet-B fuel. The smell of it was what saved our lives and immediately realized what he had done. We turned off the tank and stove and hauled him away just in time. We all stood outside at 40°C temperature getting another tent ready for me to open for a coffee break to feed our crew. Eighty four hours in a six-week stretch with twelve-hour shifts is ultimately demanding.

My military background had given me the edge to work full time with the Corps of Commissionaires doing security detail at the headquarters of the Joint Canadian Task Force in Yellowknife, Northwest Territories. I've been working in Yellowknife, one of the major mining cities of Canada's north for two great years just before this mining job offer. In the cold winter of January 2010 I drove taxi part-time on the evenings for the extra cash for me and my wife's holiday time. This one particular evening dispatch had me pick up a fellow at the Yellowknife Inn, a popular hotel in downtown Yellowknife. I could see the passenger was staggering towards my taxi, his breath reeked of booze, of course. When I hopped out of the cab and tried to put his seat belt on, I had a whiff of his clothes. Based on my experience with smells, this particular one belonged to some sort of chemical that was not common to me right away. I asked if he was OK because otherwise I could drive him to the hospital or call the ambulance. His two friends who stood by the hotel's main entrance right by the main door with a smile, saying that he was only drunk and that his address and twenty dollars was in his shirt pocket. Kindly take him to that address, which was not too far away, they asked me, as they waited for him to leave. Upon arrival at the address, I then slowly hauled him out of the car. The lady who then opened the door helped me take the drunken guy to her living room couch. The odd smell on this drunk was what I smelled on my clothing after that freak incident at the mine.

I did mention in detail to our dispatch the suspicious passenger's safe arrival at this home. The taxi office's dispatchers take notes on need-to-know information and pass them on to one another.

"Booking free "is a verbal agreement to be able to take another trip passenger or deliveries. Dispatch told me to head back as there was yet another apparently drunken citizen of his to be picked up. When I arrived at the hotel, the same two fellows stood outside as the other gentleman was being seated in one of our competitor's taxis. The staggering and smelly passenger that I had taken home on this New Year's weekend of 2010 was the same fellow I had gotten into an argument with at the mine site on the last week of January 2011. Because of the verbal exchange this bugger and I had, our German supervisor stuck us on separate days and evenings shifts at the crusher plant for the next ten days. I did reiterate the reason all that this workplace disagreement had been in defense for the Aboriginal people, me included. With my years of experience, I had accepted a nice offer in another department of this mine site for my next run. A very tiresome arduous shift on my second two-week stretch that started January 18th was finally here. But, alas! It ended with what I would consider a life-changing moment.

Before I turned 50 I was in reasonably good health with an impressive skill of being able to remember names, addresses and phone numbers in a blink of an eye, and a safe five-star driver for over thirty years and passed all of my driver training courses successfully. Now I suffer from short term memory loss and lost that impressive memory skill I had and reality finally hit me like a bad dream or a scary movie. My story will hopefully pose as an example and will show the truth if it something like this ever happens to others at a mine site or any workplace elsewhere.

Yellowknife, Northwest Territories

She said:

Day 1. Around five p.m. on February 1st, 2011, as I was walking home from the post office, I heard my phone ring. With no hurry, I grabbed my phone from my purse wondering who it might be since I don't get very many calls.

"Hello?" I asked.

"Hello Iris this is Dr. W. I am the medical officer at the Diavik diamond mine site. It's about your husband, Michael. There was an incident at work and he is unresponsive," the voice on the other end replied.

"What?! What do you mean?"

His next words sounded so distant and I wasn't hearing him at all. "He is not breathing. We are flying him out via MedEvac (short for medical evacuation). He should be in Yellowknife by eight o'clock."

Suddenly my world zoomed in on me so fast I felt I was on a 160-mph downhill roller coaster ride. In utter disbelief, questions of what-ifs swarmed my head while my eyes, surprisingly, stayed dry. I was too stunned to cry, I guess.

Hours later, I was sitting at the corner of the ER lounge, waiting. My mind still in a blur, I quietly prayed, oblivious to my surroundings. One of the off-site supervisors had offered to drive me to the hospital and keep me company as we waited. Michael was being airlifted from the mine site to Yellowknife and he was expected to arrive at around eight o'clock that night. I was anxious to see him.

Two hours later there was a commotion as they wheeled in Michael on the stretcher. It felt so unreal, as if I was watching a movie shoot. It seemed like years before a nurse came to get me and she said that Michael was stabilized. I didn't really understand what that meant at the time. I slowly followed her and then stopped in my tracks as she pushed the curtains aside and I looked at my husband. He was as white as a sheet and there were tubes everywhere. His skin felt cold and clammy. He's dead I thought, and that only the respirator was keeping him alive. I sobbed as I stared at his face covered in tubes. They allowed me only a few minutes to be with him because they had to take him to Diagnostics for a CT scan and then to ICU after that. The nurse suggested I go home and come back the following day because Michael's condition would pretty much be the same throughout the night. Obviously, sleep was the last thing on my mind that night. I communicated with friends and family to oddly find out that his sister had been called just before the three p.m. to five p.m. time frame of February

2011 from the YK. With a two hours' time difference, it was of course, later back in Ontario.

Morning came and I was at his bedside at first light. Indeed, nothing had changed as the nurse said. I phoned my supervisor that I wouldn't be coming in to work until I was certain of Michael's condition.

I said:

Under the expert care of the medical team at Stanton, I woke up astounded finding tubes and hoses everywhere. My mind was screaming "Run!" My first response was obvious. I grabbed as much as I could handle and yanked the tubes away. The monitors went "beeeeep!" and jumped off the bed and fell on the floor. A nurse ran toward me and called for help to put me back into bed. Still in a daze, they put back the oxygen mask on me and tied my arms with straps to prevent myself from getting hurt. That night when my wife arrived, the ICU nurses advised her that I woke up earlier but they had to sedate me for my own good. And my wife learned of the escape incident I attempted. On the third day of being under the powers of a sedative, the medical staff at the ICU waited until I gradually awakened as the sedative waned off my system.

As I woke up, a nice lady with a bright smile stood right by my bedside. The nurse asked me a few questions as to the five W's (who, what, where, when, and why). Surprisingly, I couldn't answer her. Nothing popped into my mind at all. When she asked who this lovely lady was, I just shook my head. I did not know her name. I'm sure glad there was someone along my side, I said to myself. When I was told that she was my loving wife, a familiar trust fell right into place like a puzzle. My wife had been there with me since day one, I was told. Praise be to our Lord!

After the fifth day at intensive care, they decided I was ready to be moved to a semi-private room. They parked me in a shared room with another guy who also suffered a massive stroke but had been there months before. I truly appreciated my friends who came to visit for it made me feel valued. I did get a lot of rest out of the deal I told them.

Day 4. Nothing much happened from the time I arrived until today. After three days of unconsciousness, I was slowly waking up drooling on the right side once again in a strange place with unfamiliar faces was a panic trigger for sure. Sharp chest pains on my right as I leaned over and noticed my toes were pointed outwards and my skin almost blended with the white sheets. I yanked away all the hoses and tubes from my body as I tried to get up and leave. I fell on the floor with a loud thump. The monitor went beep-beep-beep so the nurses quickly came in hauled me back into bed and strapped my arms down so it wouldn't happen again. The nurse emphasized that the only reason my arms were tied down was to prevent any falls and hurting myself in the process.

Later that day as disoriented and incoherent as I was, smiled when my wife, Iris, approached me. The nurse told her what had happened as she walked towards my bed then asked me if I recognized this lady. She took my smile and confused look for a 'No. But being awake and breathing on my own without the oxygen at least now meant that my prognosis was better, however, they still kept me in ICU for a few more days. What the hell was I doing in here in the first place?

Day 5. I only remembered my name because the nurse kept telling me. I didn't even know my wife's name, but I now knew

I was married to her. It's funny, though, that I remembered my parents' names in questions asked, but not those of my siblings. I had short-term memory loss, we were told, due to the swelling in my brain. It may be temporary, I was told, but there was no guarantee. Medical team took very good care of me, very professionally on their eight-hour shifts.

Did you ever experience thinking of a word but blurting out something different? Understanding spoken or written words, including the processing of information is near impossible. I would get entangled with intercommunication very often. People just don't know or don't understand that it takes me twice as long than regular people to process lots of, information that I hear and to think and express my thoughts. Most often than not, I nicely ask the other person to repeat what s/he just said. To make things worse, I get easily distracted, which only increases my stress level. It seems to me that people are talking too fast especially during phone conversations. I wonder if I could have an IVR or interactive voice response service installed in my phone that would say something like "The person you are calling has aphasia and requests that you to speak slowly" It is so frustrating and unfair. My wife knows about this part of me only too well because I tell her all the time to slow down. It's easy to conclude that nothing is wrong with me because I look physically fine. When I start to open my mouth, that's where my disabilities show, or repeat as I look for my note pad these days.

"There's a table there. Bring it here so you can sit," I said to Iris. She looked to where I was pointing. I had meant to say "chair" but instead I said "table". I asked her to hand me my "comb" when I meant "toothbrush". For normal people it sounds funny but that's how my brain worked. I have improved since then, but I still grapple with choosing the right word so I sometimes fabricate them. Sometimes "big" words confuse me and it sends me running to the computer to Google what they mean.

Prior to this incident, I was exceptionally good with memorizing names, addresses, phone numbers, and past events without a daily note pad use. They were skills that were very handy when I was in the taxi business. Short-term memory loss is my Goliath. Aphasia or word finding difficulty, in layman's terms is another limitation that greatly affects how I interact with people around me. That's where my notepad and pen did come in handy. Many folks I speak to tell me that there are many others that have forgetfulness as well. For sure not like this!

I scribble notes and mark the calendar for appointments and activities. My wife uses technology for recurring reminders to take my pills because it happened more than once that I nearly double-dosed on that 6 Dilantin per day and at another instant I couldn't even remember if I had taken them or not. Repetition works for me just as practice makes perfect.

I shared a semi -private room with a co-worker who had been in bed for a few months from a very serious stroke and he was paralysed on one side His kids came in for visits. One day I unintentionally wore his winter jacket of a cold day as my wife led me to the bank to give her access to all our accounts to insure that bills, for this taxes year was all paid for.

For a few weeks after my release from Stanton Hospital in Yellowknife, I had to see the therapists that I'd been seeing while I was still confined for those twenty-eight days. My wife observed that on two occasions, preceding my seizures, I was or had been to the hospital for therapy sessions. As she would conclude, the anxiety attacks were triggered by the sight of or by being in the hospital. I try not to dwell on the thought that people treat me a little different after they've seen me in any of my episodes. *C'est la vie!*

She said:

The doctor's initial diagnosis was cardiovascular attack or stroke but they were still giving him three daily Dilantins, an anti-epileptic drug, because he still had occasional seizures. This was all medically marked on paper.

"Does Michael have any history of stroke?" the doctor asked.

"No." I replied with conviction.

"Seizures?"

"No. He quit smoking over ten years ago and only drinks alcohol occasionally. He's a blood donor for the Canadian Blood Services and had three full medicals very recently with no red flags." These all included complete blood work as part of a work requirement in Canada.

"Did he complain lately of any health issues? Or that he wasn't feeling well?"

"No, the last time we talked on the phone, he told me he was doing well except for not getting enough sleep. On his ninth day he was still adjusting to this twelve-hour-shift of its four thirty a.m. Wake; to six a.m. start. Within less than one year he had three full medicals, two for job requirements of Ontario, the most recent one was just a month ago here in Yellowknife. All the results came back that he was healthy."

While visiting our family in Ontario doing some indoor repairs for a new tenant at our house in North Bay, ON. Michael and I had both applied for in-town work that required medicals; one of his had been just six months before as a P.S.W. for the disabled with a five-star safe-driving record. The third one was for a Special Constable position and included a one mile run in better than 15 minutes. In the fall of 2010, all the blood work in anticipation of having a child was good. Insurance coverage of all aspects came with the work he had under his belt, there was never any denial of it for a car, health aspects of his workplaces till now.

"Michael had a blood clot in the lower right side of his brain," the doctor explained pointing where he said the blockage was. "But we will do more tests and send them to Edmonton for a second opinion"

I came to see Michael before and after work. With temperatures averaging -28°C outside, I would painstakingly make my way through the almost knee-deep snow to the hospital. Since I only had my beginner's permit, I couldn't drive and taking a taxi was too expensive on a daily basis. It took me twice as long to get there with the literally freezing cold and sometimes unploughed streets.

Every morning the nurse on duty would check Michael's neurological state and asked the same questions: "Where are you? What day is it today? What month? What year?" Michael showed signs of improvement as the days went by. Since Michael proved to be out of critical condition, they moved him to a semi-private ward. That was a bit of good news although it didn't do much for his current condition.

I said:

The doctors were nice enough and the food was okay. But a good night's sleep was an issue. I complained all the time about the guy on the bed next to me, who happened to be working with the same company as me and also had a stroke, but he was paralysed on his right side. This patient was also on a lot of medication while in recovery he liked to watch sports and the news. His T.V. was very loud and it stayed on all night whether he was awake or not and I had great difficulty getting a decent sleep. Found out later I snored a little loudly at times myself. When it got too much for me to handle, I rang for the nurse to turn it off. Even ear-plugs did not work. To make my nights more challenging, a nurse would come in to wake me up for my medication just as I was dozing off. Poked on both arms for a said blood tests...thirty times-that was just too much for me to take- Get me out. "I just want to go home and get a good night's sleep," I often complained to Iris.

"I'll see what we can do about that," she replied.

She said:

One early Saturday morning, I cheated and drove to the hospital with black decaf coffee and a muffin in my hand for Michael. He was glad to see me, more so with his favourite coffee from Tim Horton's. I could tell he was restless so I asked him if he would like to go for a drive if the nurse would allow him. He was pleased to hear that. The nurse gave us a two-hour pass and Michael asked me to drive him straight home. He just wanted some peace and quiet. I gave his feet a soaking and a good rub and tucked him to bed. He slept and before the two hours was up, I drove him back to the hospital. I wasn't sure if his limited freedom was a good idea because after he got back at the hospital, he couldn't stop wanting to leave. The doctor had me sign a waiver before he performed a spinal tap on Michael. Two weeks later, the results came back from Edmonton and a stroke was ruled out. Michael had only a viral encephalitis, which explained the seizures, loss of consciousness, disorientation and trouble using or understanding words.

I said:

As my condition was greatly improving, so was my eagerness to leave and go home to our apartment. Winning the battles of the side effects of the different medication, the lack of balance sleep issues, constant chest pain and urination was too overwhelming for my injured brain to handle. Anxiety was setting in and I detested each day that they refused to let me go. One of the main reasons I wanted out was the administration of drugs intravenously and the taking of blood samples daily. It was too much. I was poked way too many times and my arms and hands were swollen and very tender. Co -workers wife was a trainee, she had over poked me. I developed anxiety disorder and often had panic attacks, especially when the time came for my injections. I was so scared for my life that I got paranoid that somebody was out to hurt me. That was the main reason I wanted to leave without hesitation no matter what. The thirty pokes of the many trainees in both of his arms while strapped or sleeping prompted him to try to leave his hospitalization of ICU, and then later on still in recovery. He wanted to go ASAP with a smile for his medical team. No driving for for a whole year for a good reason with access to a Para- Bus at 40 below was well appreciated, taxi co - workers had the door open.

"Encephalitis? Yeah right!" Michael exclaimed.

"That is a lie both of our blood work proves it."

I lived in Yellowknife long enough to know that the gentleman at the end of the third floor hall way currently at this hospital, who smoke cigarettes outside he told me he was infected with this encephalitis. Having safely driven him like many other folks out of my taxi in Yellowknife NT., over the years. Small community the size of Yellowknife word of mouth does get around very quickly I know. The one in a million lie that was told to us by staff of the mining crew, little did they know. Being a proud blood donor for many years, awareness of the six percent -A RH my blood type. Transfusion of this blood type, that without it may cause blood clots that block of the vessels. I just wanted to get away from this continuing of a nightmare.

However, the doctors would not allow me to go because they had me on a very strict antibiotic treatment that I had to complete otherwise it could be dangerous, even fatal. I would not listen to my wife's pleadings to stay and finish the treatment. My wife felt so bad for me and yet wanted me to stay for fear of making my situation worse if I cut the treatment short. With a medical background herself, she asked the doctor for any last resort just to keep me in the hospital. The doctor agreed and inserted a central catheter (PICC) line where medication can be administered without the needle-poking because I wasn't taking another needle prick. No way! Yellowknife Campus is the natural site for the Bachelor in the Science of Nursing program, as the economy is based on mineral oil and gas, mostly of the two diamond mines since 1991 yes I was all for training others as they had been doing for years.

Relieved that there was less invasive treatment, I felt motivated to do exercises with the therapists from three disciplines Physiotherapy, occupational therapy, and recreational therapy. They took turns coming in but I never remembered their names. I was working toward readiness to use a walker, a cane and ski poles for balance.

She said:

Day 28. With slow calculated steps, Michael was up and about, anxious to leave the hospital. He, unknowingly; grabbed the other guy's jacket from the closet, since they had similar jackets issued by the same mining company. But before we could go, there was one more thing to do a huddle with all the medical professionals who handled Michael's case. Present in the meeting were the doctor, the head nurse, a physiotherapist, an occupational therapist, a recreational therapist, a speech therapist, a social worker, Michael and myself. Each staff member updated the doctor on where Michael stood in his or her respective field of specialization and about the out-patient program they had developed for him. Out-patient appointments were scheduled for the next few weeks and finally the discharge note was handed to me. Girded with a walker and me close by, Michael walked slowly out of the hospital silently swearing that he is never coming back in here again. We gave the Medical team a thank you card with a dozen flowers and a big smile.

I thought this might be the best part of it all. Michael would be home at last and that means no more visits to the hospital. But that was not the entire truth. Indeed Michael was home but that was just the start of the beginning – the beginning of our chaotic life. This took a toll on our relationship. It was a tough go, more so for Michael trying to cope with the side effects of Dilantin and living harmoniously with me. It probably was the worst times in our life as a couple. I was ready to crack.

And rightly so things were not the same. Michael's love for music disappeared. He wanted the radio turned off all the time. The only time I could turn on the TV was to check the weather. I had to adjust to his mood swings and personality changes. I thought all these changes in him were due to his stroke and seizures. Little did I know that the Dilantin was not agreeing with him at all? I wondered why he was feeling worse than he did when he was at the hospital. There had to be some explanation. Like jigsaw puzzle pieces that fell into place, we realized that his Dilantin dosage was the culprit. I felt helpless. Our Yellowknife family doctor claimed that the phenytoin (the general term for anticonvulsant drug) level in Michael's blood was too low so he doubled the dosage, of his one pill three times a day.

Michael experienced the worst side effects he could ever think of. To mention a few: tinnitus or ringing in the ears, frequent urination more than usual, nystagmus or involuntary eye movements from side to side, dizziness, nausea, splitting headaches, and dyspnea or shortness of breath after the pills took effect. As he would describe it to me, he felt like a 300-pound guy had his foot on his chest, making it very difficult to breath. I felt so bad for Michael because I didn't know what to do to ease his pain and discomfort. It was frustrating.

One day, he called his friend, †Denyse, who was in bad shape herself.

"How are you feeling Mike?"

"Like hell. I can hardly breath, it's as if something is crushing down my chest."

"Did you tell your doctor this?"

"Yes. 'It's due to the pills' is all he said but didn't do anything else."

"Why don't you try seeing my family physician? - She might be able to help."

Michael felt he needed supplemental oxygen because he wasn't breathing right. Denise had loaned him her oxygen for a couple of days and he slowly weaned off that evil Dilantin while I was away. I had suggested before that we go to the ER for a checkup, but he had refused. Although during one of his seizure attacks, my request for supplemental oxygen for Michael was denied. According to the ER doctor, Michael did not have the medical condition where oxygen is prescribed. The visit to Denyse's family doctor did not help either. We felt helpless and frustrated with nobody to help us and not knowing what to do next.

Two weeks after Michael's discharge from the hospital, I felt sudden pangs of pain under my right ribcage that increased with movement. I tried to stay still on the couch. "What in the world is this?" I wondered loudly.

Michael was in the bedroom resting and was half-listening to me. "My love, if you're in such pain, go to the hospital and have it checked."

"I'll just lie here for a while. The pain might go away."

But it didn't. I waited for another thirty minutes before I decided to go to the ER. I could hardly walk as I tried to slowly climb into the cab. The ride took an eternity. I patiently waited for my turn at the ER lounge. Unexpectedly, I ended up spending one week in the hospital in Yellowknife and another week in the Royal Alexandra Hospital in Edmonton, Alberta.

"The doctor is sending me to Edmonton for further testing because they can't figure out what's causing the on-and-off fever," I told Michael one day over the phone.

"That's ridiculous! Just tell them to let you go home. Tell them your husband is sick and can't be left alone."

"I did and they won't because if I may be carrying a virus that is contagious, they will be held responsible for not containing the virus from spreading."

That was the worst two weeks ever. I wasn't prepared to be admitted to the hospital, first, because Michael was alone at home during the most critical time of his recovery and, second, I couldn't even ask him to bring me a change of clothing since he was mentally incapable of thinking straight. With the medications he was on and their side effects, he was physically there but mentally absent. He needed the help more than I did and he indirectly blamed me for deserting him in his time of need. And that was killing me. I held on to prayer and hoping that things would get better.

As the end of the week at the Royal Alexandra Hospital was nearing, the nice lady in-charge of liaising patients made arrangements that I be discharged sooner as my husband, who was just released from the hospital for seizures, was alone at home. The doctor on duty agreed. A wave of relief overcame me as she started making my transportation arrangements.

Back home again. As I slowly climbed out of the cab, Michael was waiting out by the foyer of our apartment building. He had lost a lot of weight and his skin was so pale. My goodness, I almost didn't recognize him! As soon as the door to our apartment closed, I broke down as we hugged. I was just glad to be home and to see him walking about.

I said:

While Iris was at the hospital, I tried to fend for myself. One day I had pizza in the oven and turned on the TV while waiting. Obviously, I forgot about the pizza until the smoke alarm went off and the firefighters were knocking from outside on the glass window then on my door.

"Are you alright there?" asked the man who made it first inside the apartment.

"I just burned my supper," I replied.

"Sir, you must put your timer on. Is there anybody else here with you?"

"No, my wife is in the hospital and I was just there myself couple weeks ago."

"You shouldn't be alone in the situation you're in."

I apologized for causing the commotion but was thankful that they came. It could have been worse.

Another time, I received phone calls from a, pretending to be a counselor asking if I had any suicidal tendencies. "I stated No." and hung up on that scam artist. I was definitely looking to make my life easier as I coped with my situation solely. Communication was like a stumbling block to my recovery, I felt. I tried as best as I could to express how I felt, but somehow it was not getting across to whoever I was speaking to. It seemed like nobody was listening to me and nobody cared what I was going through. What a predicament! Next stop was at a for real counsellor of Sheppell FGI for two one-hour sessions to slowly quit taking all the silly Dillantin medication and slowly wean off by choice for a better life over a period of six weeks.

Every single day was a challenge. I still had the shower chair and transfer bench to use in the bathroom. But from the walker I progressed to using the cane. Two weeks later, I opted for the ski poles. I could see the bewildered expression on the faces of people who knew me.

"What's with the ski poles?" they often asked.

"I need them to keep myself from kissing the snow," I'd reply with a sheepish grin.

I slowly gained my strength and balance, thanks to the therapists helping me out, and finally returned all the loaned bed and bathroom equipment.

Six weeks after my discharge from the hospital, while I was waiting for my turn at the clinic for routine blood work, I felt the monster coming, but it happened so fast there wasn't enough time to warn the person sitting next to me. I fell on the floor, seizing and losing consciousness yet again. I woke up at the ER with a bloody mouth and a splitting headache that was worse than a stupid hangover. The ER doctor asked if I had forgotten to take my pills. I explained to him that I hadn't taken them yet because I had to do my blood work first. But I didn't get that far. My wife had a scared look on her face. If he was taking his pills, what was causing the seizures? She wondered. Either the pills were not working or whatever was causing the seizures had not been addressed. As laymen, our knowledge may be limited so we put our faith in the medical professionals who attend to us, hoping and praying to God that they look out for our best interests. As a support worker and a Level II first-aider, I've experienced medical emergencies before. Ironically, this time the tables were turned.

On one occasion, while giving my situation a profound, mental deliberation, I was still skeptical about contracting encephalitis. Medically along with WHMIS I was taught that chemical exposure does cause a scab, having dealt with it in the isolated regions of Canada. With no doubts clouding my mind, I contacted a private investigator to validate my theory. He told me he'd keep an eye out and he'd put his ear to the ground. I sent my work gloves and a sample of my hair for testing to both private and government agencies who might be able to help shed light on what exactly happened to me in that isolated mine site just before Groundhog Day, on the first of February in 2011.

I paid a visit to the RCMP station in Yellowknife and told them my side of the story. Odd as it may seem, I gave out details, names, places, and events that I thought might be a related to the incident I'd had. Safety factor was the priority I was after so the power of a pen or the use of a computer it does not quietly cover it all up as if it never happens. This tactic of exposure will help prevent it from ever happening to others. A nice lady geologist of Ontario I had worked with the summer of 1991 for Giant mines of Yellowknife NT, 50 miles south of the existing Lupin mine site of that time. At this isolated northern site we were there for five weeks looking for gold just before that awful strike was about to take place. I spoke to her the morning before we flew back to Yellowknife NT she told me and all our staff there was no gold here. All of us were invited to an underground tour ride the following day, she was the only one that went. The others were flying south and *MÂS'KÉG* MIKE spent the night at the Gold Range bar to later received a call with a few folks at home heard that she had passed away, crushed by two tons of rock on her visit to Giant underground mine site. Prayed for her to our Lord, thankful I did not go.

I drove taxi during that God awful strike, and had met all the men who died there as well.

She said:

I personally think it was unprofessional that no doctor in Yellowknife suggested that Michael try a different drug for his seizures, knowing what he was going through. With nowhere to turn to but an out-dated but still usable medical encyclopedia by

Reader's Digest, Michael decided to slowly wean himself from the pills with his family doctor's knowledge. With research, Michael learned that quitting Dilantin cold turkey style was dangerous in itself. He was referred to another wellness counsellor in town after he weaned himself from the pills, but after the first session he didn't feel like going back. Three months after his discharge from the hospital, Michael was completely off Dilantin. He said he felt better and I could see the change in him. We sold his car to a good friend of his as we flew out in the first week of June 2011. The side effects slowly disappeared and Michael was seizure-free for a total of eighteen months.

We had endured three more bitter months before Michael convinced me to leave my full time bank job in Yellowknife, NT and go back to North Bay, ON where we owned a house. He felt confident that things would be better if we started fresh with his family close by, and also seeing different doctors with a second opinion was sure worth the while. Sadly, upon our arrival, we learned that his dad had been diagnosed with a very serious bladder cancer, a month or so after Micheal's hospitalization. This was yet another good reason to be close by, should anything happen to his dad. We had owned a house going on seven years, right by the waterfront of North Bay, near a Roman Catholic Church, with a bike path a hundred feet away. Michael felt good about this decision, and he was right.

Paperwork-wise, well, I almost perfected the filling out and filing of forms. With constant communication between Michael's insurance carrier and other offices, we were never out of paperwork to do. At one point, we thought Michael was eligible for Critical Illness Benefits but apparently not. Even the Workers' Compensation Board denied his claim for the reason that what

Michael had was not a workplace accident but only one of a medical nature. Heck, who was going to help us? Michael's injured brain decided to seek legal advice. We gathered documents that we thought would append to his claim. It still is an open claim at the time of this writing. Investigation of all names submitted of this never happening to others, flights to Edmonton, calls on phones of those who have been given the power of the pen, with a promotion as a bonus.

North Bay

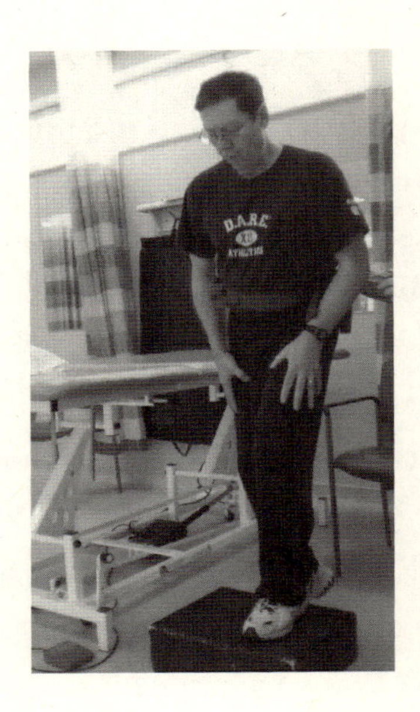

A day at the CBI Physiotherapy clinic, North Bay, ON

I said:

My employer nicely consented to my request to move to North Bay while in recovery. With a hopeful mindset, I decided to sell the car to a good friend. The three thousand-mile drive back to

North Bay was too much for my wife to do alone because my driver's license had been suspended due to good medical reasons, to find out later it had unexpectedly sky-rocketed my car insurance premium, along with many other things.

Confident that recovery would be much better in North Bay with its newly upgraded regional hospital, I sure was glad to be back home and with family close by. As soon as we were settled in our humble abode, we went to the nearest walk-in clinic and there we met our current family doctor, who sent me to a neurologist for a chance at getting my driver's license back. What a big difference it is to be cared for by a doctor who listens to what you have to say and truly looks out for your health. I had my wife's blood tested as well for any infections, but it turned out negative for any encephalitis before or after, as she had been checked before by the same family doctors. The large scab on my back removed while in ICU for five days proves of a chemical exposure, at this mine site that I worked at.

I was referred to a neurologist for further assessment before and after one full year of being without any seizures or any medication, in accordance with the Ministry of Transportation regulation. Required are diagnostic and complete blood workups before and after this whole processing. The neurologist also suggested that I try a different drug that had milder side effects, but I declined at this point, due to the traumatic side effects I had suffered in the past from the four months while on Dilantin medication. A year had gone by since my last visit to my neurologist, and he contacted the Ministry of Transportation to reinstate my driver's license. Yes I received that paper by mail in August 2012.

I was looking forward to it but it wasn't going to happen. That same week of August 2012, I had a seizure right in the parking lot of our grocery store, just after shopping with my brother-in-law, Jerry. It had been a weekend to celebrate my loving wife getting a permanent, full-time status at work. This was only the first of six more episodes: two of which were witnessed by my wife, and a few others in downtown North Bay. Some people may think that perhaps I was just faking these tense moments, but none of them are ever planned. Not just little seizures, they are for sure the Grand Mal serious falls, kicking around and biting the tongue, bleeding on my chin with deep cuts that took a while to heal. I tried to keep my mouth clean by rinsing my mouth after brushing and using a mouth guard from the dental clinic to prevent the grinding of my teeth during sleep. Many other ways of caring about one's self I learned from others through the March of Dimes.

My neurologist strongly ordered me not to drive or operate vehicles, no swimming or soaking in the tub, no handling of heavy equipment or machinery, no climbing ladders, and no heights, for a reason. Driving with no license is illegal and it had an expensive consequence. Even my wife said she herself would report me to the Ministry of Transportation (MTO) if I decided to challenge the law. I didn't have the courage to endanger myself and the public or to pay the hefty fine of at least $5,000 and jeopardize getting back my license forever. No way. So for three years, friends or public transportation were my way of getting around when my wife was at work.

She said:

While Michael was released from the ER after the sixth episode, he finally overcame his fear and trauma of Dilantin and decided to give Tegretol a try. That was music to my ears! I wished he could have done it sooner, but this was a good start. I thought Michael had enough of getting rides on the ambulance so he'd finally agreed to try medicating once again. Giving up his stubbornness paid off. They had told him while up in Yellowknife that he would be able to go without medication someday.

Even when Michael's driver's license was on medical suspension, he'd bought a nice, used Toyota Camry with his credit card, thinking it was a Corolla. No big deal. It was in great shape. The salesman drove it over to park it at our house for a surprise for me. I worked hard to get my Class G driver's license as I only had my student permit. It was such a blessing when I finally had my Class G in hand. It made things a little bit easier for Michael. Although we had a shock to hear that our automobile insurance premium had sky-rocketed because I was a new driver and it was my very first insurance.

I said:

Family and friends jokingly asked me if I smelled burnt toast just before I passed out. I told them my story the way I saw it. I sure had a hard time getting over the 'how' and 'why' of it all. Through my health insurance provider I was referred to see a vocational counsellor to check if I was ready to return to work. It took me a whole day to answer a 250-item questionnaire that didn't make much sense to me. I thought that was ridiculous. A week or so later, I was called back to the wellness clinic. The psychiatrist informed me that the results would be forwarded to my case manager. According to the letter I received in the mail a few days later, based on the psychiatrist who assessed me, my case manager said I was not ready to go back to work. Surely, not in the state of mind I was in.

On November 2011 of the same year, my father lost the battle with cancer of the bladder. It was a tough year for me and my family. But life had to go on, eh? Friends coupled with prayer, walks and Tai chi sure helped my breathing as well as my balance.

The last seizure episode I had was on August 6th, 2013. It was a beautiful day for a walk, I thought. So I grabbed the dog's leash and off we were. Less than a block away from our house, I felt the attack coming but it came on too fast before I could warn somebody in sight or call 911 myself. Apparently our neighbour, who has an auto shop a hundred feet away, saw me thrashing on the ground as he was outside working on a vehicle. He ran to me and cradled my head until the medics arrived. Our neighbour said I still had the dog's leash clasped in my hand. He took our dog with him and left a note on our front door for my wife. Just after four o'clock, my wife got home and was met with a surprise. Another neighbour met her and told her I'd been raced to ER because I'd had an attack. She also found the note on our door and went to retrieve our dog from our Good Samaritan neighbour. She drove to the hospital as soon as the dog was safe inside the house and found me awake and lying in bed. She was aghast when she learned that it had happened five hours ago and she didn't even get a phone call advising of what happened. We added her work number onto my bracelet.

"How come nobody called to inform me?" Her furrowed eyebrows said it all. "What if you had died in the event? Man that is ridiculous! Yes, they called the house, but of course nobody was home because I was at work! How come they didn't try my cellphone or my work number?" I made sure I kept important phone numbers in my wallet.

I reminded her that my Medic Alert bracelet needed her workplace phone number. Oh well. Uh-oh, she was a little upset. Finally we called up the hospital to update all our information in case of emergency.

A couple days later, she filed a complaint with the Patient Advocate officer of the North Bay Regional Hospital. I'm sure they received it but we never heard from them regarding it.

After six months of being on anti-epileptic medication and being seizure-free, I was eligible to apply for reinstatement of my driver's license. My neurologist again forwarded his request to the Ministry and four weeks later I received the confirmation from the Ministry that it was official. All of the doctor's paper work had to be submitted to the Yellowknife license bureau before as my last driver's license was of the Northwest Territories and they had a difference in regulations than Ontario. On that very same day we headed to the licensing bureau and got my temporary class G license. I was so ecstatic that I couldn't hold off from taking the wheel. Freedom finally!

My medical condition has taught me to accept new truths about myself. First, it is okay to be mentally and physically slow. Second, there are others out there in the same boat as me therefore I am not alone. Third, when prohibited by law to drive, walking and taking public transportation is the next best thing when my wife was not available to drive me. Although on a few occasions I inadvertently hopped on the wrong bus for a long ride back to the start. Fourth, follow a routine to help me remember. This is how my day begins: I have my cup of black coffee then my seizure pills with breakfast. My wife and I say our morning prayer then we check the calendar for any activities or appointments for the day. Before leaving, we make sure to have everything, especially my pills, a charged up cell phone, water bottle, and notepad with a pen should I be away from the house for a few hours. I wear my Medic Alert bracelet at all times which has information on my medical conditions and medication that may be accessed from all around the world.

I sleep with a mouth guard to prevent grinding of my teeth and with the continuous positive airway pressure (CPAP) machine for sleep apnea of a better sleep quality for me and my wife. We exchange back and foot massages in between our visits to the professional masseuse. I have semi-regular appointments scheduled with the chiropractor for wellness' sake. I ordered a Medic Alert bracelet for the convenience of the medics who attend to me during my attacks. As part of coping, I try to follow a routine and stick to it. As much as possible, I try to put things in the same place so as not to forget. There is a reminder board that has two sides where I write down reminders on the white board side and put notes, tickets and what not on the peg board on the other side of it. I even have a list of phone numbers just in case.

As a proud member of the Royal Canadian Legion Branch #23, I attended informative sessions with other stroke or brain injury survivors every first Wednesday of the month. This support group offers helpful ideas with coping for the survivors and their caregivers through the Brain Injury Association of North Bay and bi-weekly group meetings with the Ontario March of Dimes. During one of these get-together sessions, my friends encouraged me to get back on the pills. Like others whose medication is on a daily basis, some work well and others do not. While for me, a different type of pill called Tegratol finally did work. With encouragement from my wife, family, and friends, that's now over one year of doing so.

Having cooking experience over the years, I sometimes participate helping out in the kitchen during activities with the Knights of Columbus at the Holy Name of Jesus Parish. I volunteer with my wife along my side during activities involving work in the kitchen. To continue with living a reasonably healthy lifestyle, I went to the outpatient department of the North Bay hospital for physiotherapy and occupational therapy for a good six weeks and later enrolled in a Tai Chi class that is very helpful for breathing and my balance. I try to make healthy food choices and enjoy daily walks, whether alone or with my wife and our Jack Russell-Chihuahua, Chance. I am grateful that I am still able to hop on my bicycle now and again but I make sure I wear my helmet and my medic alert bracelet. I used to hold a Class F driver's license which I lost after it was suspended. The Class G is good enough for me.

Not the driver growing up, I had been in a few car crashes and survived them all. In the workplace I can proudly say I have been careful with handling equipment and only had two minor slips and fell over once when I was going down a flight of stairs with a sack of sugar on my back. There was no warning signage that the floor was just mopped and was wet and slippery. As my load was heavy, I didn't see the wet floor on that stairway. My back was so badly bruised that the following day I was bent like an arrow and couldn't stand straight for a month. What a relief it was when the doctor said I did not break any bones, but I had to slowly stretch and see a chiropractor for the next couple weeks.

The other fateful incident was when, in one of my bush runs, I was fishing on a cold misty morning. While walking a few miles back from the helicopter drop - off at that secret spot seen from the air and was a good fishing ground, I lost my footing on the moist, slippery rock covered with moss. I quickly lost my balance and ended up at the bottom of a twelve-foot rock cliff. With ripped jeans and a stiff right bloody shin, I limped back a quarter mile. I contacted by radio a medical practitioner who gave me verbal radio confirmation that my leg was not broken, then with the help of a co-worker I applied a disinfectant from the first aid kit and wrapped it with a pressure bandage. With a crutch I safely walked around until the pain and swelling subsided. I sure was lucky I did not break my leg, I later got checked up by our doctor at the walk-in clinic in Yellowknife, NT. My honey and wife, the love of my life now for almost eight years now, praise be to God, she's there to double-check on the daily process of everyday life.

This being said, I am just thankful that I have a wife who stands by me. Having driven public service in three provinces, I have seen others who live alone and fend for themselves during recovery. The March of Dimes along with others help folks who need it.

Though my body function, while on six pills a day, is surely not what it used to be, I am just thankful to God to be alive.

CHAPTER 2

MY KINS

My parents with sister, my brothers 1963

The Place Close to My Heart

Bonfield, ON, Canada is part of the Nipissing District and is on its river system leading to the Great Lakes. This is where it all began. It is a known historical fact that Bonfield is the first spike location of the Canadian Pacific Railway (CPR). Of the 208.43km^2 (80.48sq mi) total land area of the township, my parents lived on a 100-acre property with a small farm along Highway 17 (Trans-Canada Highway), which was bestowed to my father as a wedding present from his parents. On this lot, my parents built a two-bedroom house, a 50x30 barn, a hog pen, and a chicken coop. The farm was the main source of meat. Root crops came from in-town or local gardens as my parents haggled with trade to relatives for meat in the fall harvest. My grandmothers brother lived a short distance away with a huge garden we would be able to pick at any time during the harvest along with helping to prepare for the following years harvest loading his horse Chub with the trailer with some manure fertilizer for the garden.

My First Peek at Life

One Sunday in the fall of 1958, Mom was already up even before the cock crowed. The cramps she felt in her belly told her it was time to go. *Too early for a Sunday drive,* she thought, but off she and Dad went to the North Bay hospital to welcome their first born, me.

I was six months old

Five years later, I became a big brother. I had two brothers and a sister within a few years. My paternal grandparents lived close by so they helped with the chores at the farm while Dad went to work at the lumber mill and Mom was caring for her little ones. As it is in small communities, relatives were always there to lend a hand when needed. When I was old enough, I helped around the house and the barn while my grandmother babysat my siblings. In the winter, I would help carry water for the few laying hens, a goose or two, and of course the cattle from our well, which was a short distance from the barn. A big block of salt nearby to assure our accurate daily critter count for the hay feed, was sometimes a surprising extra one of wild game. Often had to learn helping out cleaning the barn early before school with a chance to wash. When my brother was old enough to pitch in, we would horse around while doing our chores, ride the calves, or chase the geese. It made the day fun. Sharing the workload to feed our livestock every day before heading to school on the bus, and watching our step with the dingle berries.

A Kid at School

Before my first year in school, Mom coached me on how to dress properly and mind my manners. On the bus I met a few cousins and made new friends. But I was careful not to tease or horse around or get into a scrap because we would hear rumours from our neighbours about that.

We were the last kids to be picked up by the school bus in the morning, just before 7:15. We were guaranteed a seat at the back of the bus because we had the distinct barn smell with us and the other kids didn't particularly like that. Oh well! My siblings and I all went to the *École Sacrée- Cœur* or Sacred Heart School in Corbeil, Ontario on the King's highway # 98. The township line was part of the bus service and we lived closer to it than the school in Bonfield. Our school was run by nuns and priests and a few great local teachers.

In school, the sisters taught us prayer and singing and lots of other fun stuff such as playing baseball, soccer, and track (running). After the bell rang at around 8:50, I remembered we had to kneel on our chairs every morning for prayer to start our day. Teacher would do a head count making sure everybody was present. If you missed class, you had better have a note from your parents the next day. If you got into a fight or any misdeed, you got the hot seat in the corner or you were asked to write on the blackboard:"I will not do it again." type of punishment. And like many boys, I had my share of the leather strap from the nuns for misbehaving. And I honestly admit I deserved every one of them.

If you wished to engage more in music or sports, the opportunity was there. We would perform in local events and the best one of all was singing for the seniors on special occasions at the

Nipissing Manor, which was a mansion that once had housed the Dionne quintuplets. The seniors were always happy to have kids around. I also had the chance to play the trumpet for the school band because it was the last instrument available...nobody else wanted it.

Mom's brother, Uncle Rene, told us the stories of the time he was in school. Apparently, the nuns and priests were much stricter then. My uncle was a *south paw*, he wrote with his left hand. He was forced to write with his right hand or else he would get the strap. To this day, he is ambidextrous but prefers to write with his left. This worked to his advantage when he started working as a roofer. He could switch hands when the other was stiff and needed a break.

Tenth grade from École Secondaire Catholique Algonquin was my highest formal education attained because I earned my GED diploma in 2007 - years later working days and my nose stuck in my books during nights. It always bothered me that I had quit school just too early. I realized what the job opportunities are when you don't have at least a Grade Twelve. I remember Dad always said, "If you don't finish school, you'll end up doing a job that pays minimum wage just like me. He was right. He didn't mean anything negative about minimum wage jobs. What he actually meant was that there was very little to no room for advancement without a good solid background in education, to make it happen. He had a pair of gloves with safety toe boots for the bros and I to do labour work, work for which he was well liked for more than forty years, with a spare outfit for each one of us if that was what you chose.

My grade school picture at Sacred Heart School in Corbeil, ON

Algonquin High holds fond memories for me. In my nearly two years there I had learned practical skills such as electrical, welding, and mechanical for needed use of the math skills. Two years of musical class playing the trumpet gave me a passing mark.-Guitar was my personal choice, which my father played as well a violin. My social life started to bloom with my achievement so far. Now that my exposure to a different sphere of growing up was increasing, so was my curiosity of searching for more. More of what? I wasn't sure myself. This led me to decide I wanted to earn my own money and be more independent. Bad decision. I quit school in 1973 at the age of fifteen with my grade ten, and went to work in Toronto, Ontario.

Mom

Sending off Madame Dionne (in purple coat) at the airport in North Bay, ON. A nurse (left) came with Mrs. Soper (in white coat), her kids, me (nest to the nurse) and my sister (purple sweater)

Our loving mother of Catholic upbringing trusted this nice German lady with us kids at all time as were being care for right by her side in a healthy proper way. Sure gave our father a break that he enjoyed his free time while getting our new home built. During the summer breaks before I had quit high school, our dear loving mom would often take us by owner's invitation to the Nipissing Manor where she worked full-time as a support worker for over twelve years. In 1966 Mrs. Soper, the German manager/owner had taken over the huge Dionne Quints former homestead, which was slowly expanded as it turned into a fine Nursing home it is today. She hired local staff to care for the seniors, on the King's highway 94 of Corbeil,, ON. Part of the lot was her large, nearby staff housing with a lovely homestead and lots of room for her two younger kids who were of our same ages. We would play outdoors or sometimes help assist the seniors as they went for their walks on a sunny day. At times Mrs. Soper would put us to work watering of the plants, washing walls, and yard cleaning which, of course, included sometimes walking the dogs and picking up dog poop. For a treat, Mrs. Soper would oftentimes request Junie Prunie, her second-born daughter, to haul all of us kids in the back of her station wagon and take us all out to a bowling competition, along with watching the latest movie. Junie and her husband, Stan, would take us fishing and swimming not far away in Astorville, right on Lake Nosbonsing, where her mom had a cottage. Of course our parents were invited at any given time to stop over as well.

Those fun loving summers of games, we would have a yummy barbecue feast after the swim. That is where I learned to water-ski and,-drive one of the first ski boats,- Junie's husband Stan had a few secret fishing areas for the skidoo winter option as our dad cheered us on with permission to do so. Mrs. Soper had an oldest daughter named Bridget who lived at a farmhouse out in Rutherglen. ON. Less than an hour away. Over there we got to clean up around the yard and our pay-back was to play outdoor games or, fish for trout at the nearby trout pond best overall was to ride one of the six horses. They sure had a lot of fun too! Then the feeding and cleaning up after the horses came next. Mrs. Soper treated us like we were her own kids. She had big dogs at the Manor to play with, but mostly they were for protection as the property was three and a half acres. Throughout the summers we spent time with them at the Manor and they sometimes came to our farm in Bonfield and us kids would play among the bales of hay in the barn and swinging on ropes, pretending we were Tarzan. But we made sure not to mention this to our dad or else we could get a licking.

Eric, Mrs. Soper's son-in-law, had been a professional German soccer player who had also been raised on a farm. He would teach us a few of his back-yard scoring moves. Raised on a horse farm himself, he had us with the girls safely participating on a warm summer day, setting up the practice jumps for his well -trained, dominant white stallion. Later in the day he would have us help clean his true white horse as he showed us how to tie him to his stall, followed with a clean-up comb and brush, properly feed him after this practice session. Then he'd take us all out back with one of the un-ridden horses to now be rode by us newbies. Quietly, he'd watch as we safely rode the new horse. Its harness was tied to a steel peg and rode in a circle. As the bravest one of us got to be the first to slowly hop on bare back, we'd do so with a carrot treat, softly talking to the horse then slowly climbing on it, using a bench to hop on his back, and then grabbing the reins. After many turns round the expanded circle and stopping for a break, we brought the saddle to put on the horse for that great ride. Yes, we were all sore holding on for this ride as our dad Percival came over a few times to view this awesome riding technician's ability to pass on his skill to us. We would all laugh as we told our dad some of the methods we were shown to make the horse listen a little at a time with patting soft talking.

Talking about Mrs. Soper's son in-law was an all- around man at the Nipissing seniors' home, covering the base points of many services; taking vehicles for repairs, garbage clean up after the pickup of a huge bin, or storing heavy food items to the kitchen. He was well liked by many as part of the team especially for heavy lifting. His loving wife Bridget worked the kitchen, and we got to spend time with her now and then during her weekday. She did the ordering and inventory for her mom then at her day's end give us some of her pastries.

On one exceptional occasion while the Manor was being converted into the nursing home, before- the expansion to what it is today, Mrs. Soper and her kids gave us a chance to meet two of the Quints with their mother one day. They had an original church play area at the lower part of the acreage, where many years before this, tourists would observe them behind a two- way glass. With Mrs. Dionne's approval, we were allowed to sell postcards and souvenirs of the Quints to tourists at the lower entrance -way, and then take them for a walk in tour, one group at a time. It was supervised by her son in law, good old Eric the horseman who was part of the German family as he played an important role at this Nursing home for incoming guest and all staff. They all made sure that we kids were always around an adult before heading back home or staying in for a sleep over. The Dionnes' original house was eventually moved, and is now converted into the Dionne Quints Museum. It is located in North Bay, Ontario just past the entrance-way of the Trans-Canada to Highway 11 to Toronto. Sure was a memorable time for all of us kids to participate in doing so with pride. Their sister in-law who lived next to the Manor was one of my school teachers in Corbeil, ON. Just up hwy 94 called the King's highway.

Summer – Back at the farm...

I grew up in the country on a small farm on Highway 17 East, and my family had limited access to the outside world beyond the fences of their farm, but unlimited access to our fine Canadian nature right in our backyard. I may be exaggerating about the first part but let me give you an example such as firewood used on a cold winter's night or a cool, stormy day, the radio and television were the few sources of media entertainment, if you could call it that. To dampen our spirits all the more from not being able to play outside due to bad weather, we had to be content with the one or two channels on television – CBC news, the famous "Hockey Night in Canada", soap operas, and Don Messer's "Jubilee." On our French channel that was well liked there was event news, a soap called "Les Belles Histoire des Pays d'en Haut" and a favourite kids' show entitled "Bobino". But if we were lucky and Dad pointed the rabbit ears the right way, a third channel was definitely a nice weekend bonus for us to catch a movie. Most of the time, rather than enduring the drone of dialogue coming from the soap operas, we would go outside as the weather permitted. We had acres and acres of land for our backyard. We were living off the land in a contemporary sixties style with a fresh water well that was popular in the community, and crops in the field with farm animals to boot.

As I was old enough to help out, part of my chores was to follow Dad around as he worked to familiarize myself without getting a kick from the cattle as we cut their horns and checked their hooves. We'd distribute a bale of hay to each stall, or get a milking lesson for a spring born calf till its weaning -off.

My granny watched my siblings as we fetched water in buckets from the spring well to the barn, a short distance away. We did this again in the evening. We were sure glad when my dad finally installed a automatic water pump. Mid-summer was the time when hay was harvested and bales of it were covered with poly to keep it away from moisture. They were later stored at the barn in stacks. Sometimes we got to ride a young pony or calf without a saddle. We fed it grain, apples, and carrots for a ride, using twine rope for the homemade reins or to tie it to a sleigh on a winter day.

Back then, I didn't realize the worth of the farm because for me the chores took me away from playing. However, my brothers and I loved to play in the barn even after Dad strictly forbade us: namely for the fear of fires or getting in trouble with the hogs or the geese as they had a bit of a temper themselves. Of our pet dogs with a cat now and then the back yard was a lot of fun. The geese sure bit hard, the hogs did not like to be rode, and neither did the calves as we took our turns. Dear loving mom would always have an eye on us to, making sure animals are fed.

But like most kids, the more he said "No!" the more we sneaked out there behind his back. Our neighbours' a was an Algonquin family of the Antoine's with their six kids, would come over to play "Cowboys and Indians" out in the hundred acres back yard a lot of room. With the thick ropes hanging from the beams of the barn, we would swing to the other end of the barn and yell, "Awoooo!" " just like the old days, or so we thought-With Percy our dad he let us set up the backyard with a real teepee. The head of the Antoine's household, with our dad's permission, did some trapping in our back yard and showed us how to make bows out of pliable, hard maple wood. The arrows were made of wooden sticks, sometimes from the mill, with pop caps attached to the end so it wouldn't hurt as much when it hits you. There was sling-shot playing with a target in place. "Hide-and-Seek" was another favourite game we played. The girls did a lot of skipping now and then, watched we boys play the many games.

Summer was the best time of the year for us. We never ran out of things to keep us occupied. Living close to the Mattawa River and many other lakes nearby, our uncles showed us how to hunt, trap, and skin wild game. Dad did not particularly like an unknown wild game meat we sometimes got from our trapper uncle Turcotte.- One day it was cleverly spiced by Mom who made a peppered garlic, savoury sea pie dough. He said, "Don't tell your Dad!" "Dad had a taste of it and thought it was good, not knowing he'd just eaten a good feed of wild game! Should there be any doubt he'd ask us to go get him the ketchup!

Living in the country had its own quiet and lazy days. A television set or the radio could be heard in the house. We only had one of two snowy channels on TV. On a lucky day sometimes we get a third channel if the rabbit ears were positioned at a certain angle. We, the kids, preferred to go outside and play in the barn or in the bush. If we had enough players we played soccer.- A little tag was liked but hide and go seek sure had some moments with the girls. As we had only one sister she was very competitive with our nearby cousins. The bicycle rides downhill with friends and close relatives always entertained us. Once or twice we would witness or hear about a vehicle crash along the highway. Even our dog wasn't spared from this hostile reality of living along the Trans-Canada Highway #17.

A friend of mine and I, one day, decided to go pick apples. So we hopped on our bicycles and went pedaling the gravel road to a property that my grandmother once owned. The apple trees were teeming with ripeness right to the side of the road and we decided to pick some. *The apples are falling on the ground and only rotting anyways,* I thought. So we filled up our brown paper bags as fast as we could and pedalled away. On a steep hill, riding with one arm and in no particular hurry, I lost my balance, fell off the bike to hit a rock and busted my bag of apples. I didn't realize I had nicked my head until my friend said I was bleeding. So I took off my shirt and wrapped it around my head. We came off the gravel road back onto the Trans-Canada heading home, when all of a sudden – *Suffering Succotash* – an OPP (Ontario Provincial Police) patrol appeared, hit the horn and pulled us over. We got scared because we thought we had been reported stealing apples. The officer asked if we were okay. I told him I'd had a crash and was heading to my friend's house because his mom was a nurse and she could fix me up. Turned out to be a trip to the hospital for nine stitches right at the top of the head. When I got home I told the story and gave the apples to my granny. She made us a delicious pie. God works in mysterious ways doesn't him?

Granny told me, "You come close to apple sauce instead, eh!?"

Some afternoons in the summer we would go berry picking. Mom always warned us of the wild animals – which wolves or bears could be nearby – to watch for tracks and then use our whistle or bell to make some noise. When I would hit a patch where the berries were plentiful, I would stay as quiet as I could be so as not to let the others know I'd found a good spot, and I would have the berries to myself. One day, my sister stepped on a wasp hive. We scrambled and fled as fast as we could as they followed her. She was in a lot of pain, although if I mention this incident to her now, she'd deny it.

After berry-picking, we would go to the a little spring and go for a cool dip. Once we got home, with our parents' permission, after there was enough for the pie, we would sell some of the freshly picked berries on the roadside. Mom would make lots of pies from the berries, have one for a taste and freeze the rest to later use on gatherings. They would last us through the winter when berries are expensive at the grocery stores. Sometimes we would trade one of mom's pies for a tub of ice cream at the local convenience store.

Summer time at its season, once a week we would fill up burlap sacks with *Lycopodium*, also called ground pines or creeping cedar and somebody would come buy it for eight cents a pound. It actually is a medicinal plant but many people use it for decoration.

One of our favourite aunts, Auntie Irene, would sometimes babysit us during Mom and Dad's Bingo nights or when they went shopping. She and her husband, Uncle Oscar, lived about a quarter mile away from us. Since they didn't have a car, Uncle Oscar would come pick us up on his big black Clydesdale named Chubb. We boys would climb up on his back while my sister rode on the wagon behind. "Gee-ha!" my uncle would say to Chubb and he would start walking. We would give him a pat on his side or a rub on his head to say thank you for the ride. A piece of carrot also worked. Chubb was Uncle Oscar's right hand on the farm, especially when the harvesting of hay arrived. In the summer time, we would help him out by picking vegetables from Auntie Irene's garden or bring firewood inside the house as a payback for the free babysitting she did. When nightfall came, Auntie Irene would feed us and then we would play cards or board games. As we asked her for a headband to wear, she even taught us how to knit! She told us that is the same way they make fish nets only the knit, pearl two are just bigger. Other than that, storytelling was always fun. We would hear their awesome stories of successes in life or catching up on the latest news about other relatives. Then Mom and Dad would come pick us up. Auntie Irene would haggle a roast in trade to have us bring home a bag of fresh vegetables. Our kitchen table always had fresh food such as pork, beef, chicken and eggs from our farm. Fresh vegetables we got from Auntie Irene. Only a few items were store-bought. That was life on a farm.

Before the winter came, we would help my dad saw wood for the wood stove at our house and of course at his now widowed mother's, our Granny. Although my granny was a physically tough lady, we pitched in the gathering of wood from our 100-acre backyard, and then from Granny's 100-acre property too.

We had very interesting neighbours the Antoines. They rented my granny's house for five years. They were Aboriginal Algonquin's of the Mattawa area we were said to be related. From them we experienced real hunting, fishing and fur trapping just like they have been doing for many generations. They even taught us how to make a fire, and how to create and use properly bows, arrows, sling shots, and spears. In the winter time, we learned how to wear a pair of snow shoes that Mr. Antoine would carve from scratch – along with my dad and uncles showing us a few tricks; observing the ground or snow for fresh tracks when snaring wild rabbits. And there were ice fishing, ski-doo, and sleighs.

Growing Up

Each of us has his or her own unique childhood stories. My life as a kid and later on as a teenager, in some sense, was not a bed of roses but more of a thorny bush. Our family was not dirt poor but we couldn't afford anything more than the basic necessities. Dad worked for minimum wage and that was barely enough to feed four kids and keep the farm going. Mom eventually got a job at the Nipissing Manor to help support our family. Our haircuts were done by Auntie Irene for free. She would wrap a towel around our backs, perch us on a stack of Coca Cola crates, and with her straight razor make our hair disappear. This made the girls giggle.

Dad

My dad Percy with his sister Marie and their parents great grand father Felix St-Pierre

Percy & Marie with their dad Henri Ouellette

Labour work was Dad's reminder for us that education was the way to go. He worked for a lumber mill in North Bay five to six days a week for minimum wage for ten years. Having only finished third grade, he didn't go very far with achieving a handsome career. He was friends with many and his labour work was a means of raising a family for my dear old dad. But because of his lack of education as a reminder, I sure realized later on in life that dreams are limitless if you only dare to explore.

He did the shoveling in the barn, lifting the bales of hay and checking for any critters nesting there before heading out to work. In the winter he would tell us, "If you want to earn your quarter, you'd better get up and grab your shovel before Granny beats you to it." True enough- Some days my granny would be up before the rooster crowed. By the time we walked out the door, the shovelling of the driveway was halfway done. She was a tough, hardworking lady and could easily beat many men who were willing to challenge her in arm wrestling. Dad always said, "Never mess with your granny."

In his spare time, Dad would take out his violin and play some tunes while seated on his rocking chair. We would sit close by with a harmonica and a pair of spoons with humming and whistling to accompany the tune. Mom would be in the kitchen punching out the dough for her famous homemade bread with some pie dough left over used as a *pet de soeurs* aka nun's farts, which was a well-liked rolled pastry with butter, brown sugar, and cinnamon cut into half-inch pieces to be baked.

On one of his hunting trips, Dad and our uncle took me with them. This was my first duck hunt using the dress code of greenery and the quack methods to attract them for a good shot. Then a walk around the swamp looking for the eggs in a duck's nest. As we had caught two ducks for supper, Dad had us sneak the eggs under one of the laying hens. When the eggs hatched the mother hen did not mind the three ducklings at all. It was funny seeing the ducklings go for a swim in the creek while the mother hen was walking up and down the water's edge clucking at them to come back. That was one of the stories my parents would tell visiting family to just go out back to check it out if you do not believe us.

Dad would wake us up at five o'clock in the morning each day to eat breakfast and get ready for school. He would have stacks of toast or bowls of porridge ready on the table. Then off he went outside to the barn to kick-start the day. He would chat us up on how to stack the wood correctly into the stove for that winter's morning. Later in the day after school, we would help stack up some more wood until I'll turned ten years of age when we did get a diesel powered furnace with wood for a back-up. For many generations the wood stove had been a means of lifestyle all year round for many. Still is for some to this day.

CHAPTER 3

MY LINEAGE

Library registry of marriages, and of our over 400 years in Canada and the U.S.

Paul J. Bunnell, FACQ VE "French and Native North American Marriages of 1600 – 1800"

"Repetoire De Mariage du Moyen Nord" all nationality listed volume **8**- Bonfield 1882

Mattawa 1863 Lucien Rivet c.s.v. 1125 "Mariages du Comte de Terre Bonne" Montreal 1972.

"Mariages du comte des Deux Montages" Montreal 1970.

Rosaro Gauthier "Mariages de la Paroise" St- Laurent/Montreal 1720 -1974 (Bergeron).

Rene Jette "Dictionaire Genealogique des Famille's du Quebec" 1730.

Dictionnaire Biographic du Canada vol. 2 1701 to 1740 Laval Quebec "Chief Madokawando"

L'Abbe Cyprien Tanguay vol. 3 "Dictionnaire Genealogique"Familles Canadiennes Laval, QC.

Like the Kings and Queens of other parts of the world married into one an-others families to make peace, for all aspects of life and trade, so did the Aboriginals from the start as the collective public mind, does fade quickly, even after a quarter century was there for a good reason. Algonquin Treaty negotiations of Ontario is constitutionally protected from the start.

After two hundred and forty years

My great grand father Felix St-Pierre daughter Anastasia and Melina.

Here is a picture of both my gr-great grand fathers Michel Ouellette & Adelard Leblond displaying an Algonquin lifestyle off the Kings hwy 94 Corbeil ON Canada I was told, just a hundred years before now. Adelard was a security guard at the entrance way in early 1900's.

Raoul (father) & Henri (son) Ouellette

Basile Adelard Leblond & Henriette Elisabeth Blondin and their daughter Evodie

My grandparents Henri & Melina Ouellette. 1955

My great-grandmother Rose Drapeau Amyot (with the maple leaf)

Evodie Leblond b.1889

Saint Pierre et Miquelon of the colonial empire, is still owned by France. A new Suzerainty definition Exemplary start of Jan. 29, 1712 the **Peace of Utrecht**, a historic series of individual peace treaties. The agreements of peace and trade for land and water to bless the fishing, fur, lumber, and mining industries for all. These rules and regulations still apply today coast to coast of Canada because of this agreement. Aboriginal are all part of it from the start to live in peace to enjoy life in civilized way working with one another in the profits of the agreements.

In the wake of the Industrial Revolution, a natural abundance was exploited by work resources of Canada. My relations on both sides of our family are descendants to well-known explorer Samuel de Champlain, his translator Jean Nicolet as they are both left in memory of their achievements in writing on monuments in many places of both Canada and the US. There are a few in North Bay Ontario. Raising large families this was all part of survival in the long winters, trapping for food of course to sell the furs at a later time to all be sent back to European section by ship.

With a monument and plaque in honour as a Canadian carpenter, Guillaume Couture who came from France, had helped the Jesuits using his learned translation skills of six native languages to be part of this settlement. Levis, QC. Which is just across the St. Lawrence a short distance from Sillery QC the First Nation Reserve of Canada. Guillaume's education with major trust in place put his legal ability to use and it played a big role in the court processing of that time. He was given the brave military rank of Captain for the militia of his area so he could adequately do so by rank. From the 1640's to 1701, Guillaume Couture with his loving wife raised ten kids while living in Levis, QC. Till the age of 82, he efficiently put his time in as diplomat, where still to this day, our relations are part of this travel route to Ottawa.

In 1857 Queen Victoria was asked by her advisers to choose Ottawa's location as the Capital of Canada.

Ottawa was a French and Irish community that had started back in 1855 with the name "Ottawa" that is derived from the *adawe*, meaning "to trade." The Grand River was a secure, practical route traveled by many as it was right in-between Toronto and Quebec where the bridge and then the railway connected to New York in the United States. Now the fairness of haggling trade has somewhat settled in many aspects, through the years. As the bidding of furs is still an open house sale at various locations in Canada till now. Team style control of manipulative payment for all the skunk furs that is more well-liked by the Russians as I did get to experience it on a spring sale as a fur hauling helper.

The historic land claim of the Algonquin Treaty is finally falling in place involving their kids, language, and their family's right to hunt and harvest the land to barter for cash for survival. Algonquins are still here. British and Canadian authorities recognized that indigenous peoples already on the lands had a prior claim to its Aboriginal title. The Royal Proclamation of 1763 and the French and Indian War did stabilize the Europe, Canada, and US relationships to ensure the legal significance for **all** the First Nations of Canada. Our identity, along with land claim negotiations, was agreed through inherent rights by policy from the first day on, by factual history of our descendants through a well-organized lineage documentation, and now DNA. European descendants married Aboriginals as a part of life.

Many Canadians registered legality processes are based on past events and currently used as evidence and in proof. Though there's an expectation of honesty from all our "Chiefs" and leaders, to pursue our rights with integrity to our best interests, treaty does still come first in today's society. Equality for all members as an ON. Algonquin's two hundred-forty year old appeal that is finally on its way. This is for real, truly part of our society.

Published by Highway Book Shop of Colbalt, ON under direction of Major G.L. Cassidy, "***Warpath***". Dedicated to the Algonquin Regiment WWII 1939 -1945, second printing 2003. We have direct lineage to relatives who served proudly, which was officially registered, bringing our levels of life to what it is today. Aboriginal's tradition of burying the hatchet as opposed to the swastika legends of its mistakenly twisted history.

Neh-ka-ne-tah: "We Lead Others Follow"

Nicknamed, Algoons Gonks, Meegwetch to the Great Spirit, Aasha–Kitchi Manitou, our wider group.

Ontario has a regulation in recognition of the Metis lineage with its proof of being a direct descendant within five of those generations who ruled in the easterly directions of Manitoba, Saskatchewan, Alberta, NWT, and British Columbia. As the Metis had scattered all across Canada and the United States they essentially coalesced into a one Métis tradition. Although like many others *MÂS'KÉG MIKE* being from Ontario was past the five generation government ruling in being a Métis. Firstborn and light-skinned, in elementary school I was asked by few others if I was adopted as my sister and brothers were darker-skinned than I was. Like so many others in the Nipissing District of Bonfield, ON, we were brought up silently knowing our Métis cultural ancestry with related family members still living close to one another from the start. Before completing their grade twelve, many moved away for work, hitchhiked to Toronto, joined the military or ended up in jail as their first mistake in the outer parts. There are Ouellettes living in all other parts of Canada. I was accepted as a Métis in Quebec, North West Territories, and Nova Scotia with proof and my family lineage name.

Regulations have slowly changed for a reason; an Algonquin family relation from Mattawa, a nearby community to Bonfield, was taken off their property due to this unfair treatment, like many others who were just told to leave. What if an ex-spouse or child would be just short of one day to qualify under those silly, manipulative rules about that time? Our own relatives wrapped a wire fence around our one-acre yard in the silliness of stating we had Aboriginal blood, so we should just be pinned down, Ontario Provincial Police stopped them. OPP made them dig out all the post, then apologize to us as we had then enough lumber to finish building our back yard teepee.

Métis means a person who self-identifies as one him or herself and it is for Canadian Aboriginal peoples, of historic Métis Nation ancestry to then be accepted by the Métis Nation of their. In the legacy of racism, the rights of women raised on a reserve were lost if they married a white men. Aboriginal ladies married a white man would lose their rights to being part of a Treaty, as their kids now became part of the Métis lineage. Their children have now sensibly been re-issued there rights as a rule seems to be a onetime choice of with who do you want to be with. There's a legal process of disenfranchisement to terminate a person's "Indian status" with a blinding quick signature. Voting has now come a long way here in Canada. Many family members stayed close to one another, but many did not, due to the confusion of marital status regulations. Yet some were taken in as leaders without even being of Aboriginal descent. Tightly knit families and cultural traditions were kept closely linked. Franco families often gathered for soirées, where people would talk, tell stories, sing, and cook. Lumber was an important source of wealth. In 1985 the Indian Act was amended to restore status for those who had lost it through time for all their children because of discriminatory provisions of the act.

In 1772, the 7,000 folks of our Canadian population, my ancestors, were part of this event. The lineage still exists, along with us still living in this area since that time. Algonquin's unsurpassed peace treaty did survive, to this day. Now the first modern-day, constitutionally-protected treaty for the Algonquin's of Ontario, Canada, is about to finally take place after twenty-four0 years. The motto of "We Lead, Others Follow" is finally being respected for doing so. Bravo! Hear hear! ... To finally pound the gravel and punctuate the rulings and proclamations.

Like many other political issues, over time an election puts things in perspective for the voters. In process is the Quebec Act of 1774, where petitions had been registered dating back to

1772 asserting the Algonquin's' rights to their lands for the future. Treaty rights and settlements have fallen in place for the Algonquin descendants, for their children's future.

Where the rights of cultural minorities like the Algonquin's was in confusion forbidden to conduct or rather more so ignored as characteristic in ethnicity. A regulation of losing after 5 generations of even to being a Métis was a quick vote passed without asking.

Where both sides ratify a mutually desired agreement, to imply recognition of our Canadians' then-government's role in achieving realistic goals of the past and now for this day and age.

With time, education, and now our worldwide technology of communication, a better achievement potential has been brought forth for us all. Online voting in the election process – modern technology, eh!

Referenced section 25 the Constitution Act, of 1982. Labelled "Indian Bill of Rights." INAC. Said to be a landmark of aboriginal self-governance Nunavut Act for Claims agreement was signed May 25, 1993, was read into law on July 09, 1993 by Parliament of Canada. Became a reality in 1999. Yes I was teaching a cooking course in Kugluktuk 2000, being part of that myself.

As being their neighbours, Bill C 15 for the NWT, gave Northerners greater control of each others land and resources by the "Devolution Act." Where minerals gas and oil are still in a big demand.

I am the father of at least one Aboriginal child from that region, where I lived and worked for half of my life in Yellowknife, NT, as part of the public service as the owner/operator of a taxi business for many years. I was given a nice picture of a lovely young lady to let me know our DNA is of the same. I received a picture at the cab stand, hoping to meet her one day.

Métis registry as a prospector of a diamond mine in that region, I have shares of the stock market yes, they have sent me a diamond. Part of the agreement to become First Nation is to belong to only on governmental agreement at a time.

Today *MÂS'KÉG* MIKE is a proud current member of the Mattawa/North Bay Algonquin's First Nation of Ontario, Canada.

I am a registered Métis of QB –Yellowknife, NT region but not of Ontario because it ends after five generations. Today's society with modern DNA technology lets peoples know more in a short bit of time. The Algonquin treaty was part of all this from the start to what it is today.

Where I've defended myself in small claims court over the purchases of a crappy vehicle to be used as a public service, namely a "taxi-cab" that looked like new. A maintenance worker friend of mine reminded me that this vehicle had made the newspapers. They had not told me about this part. It was had been sold to me after hitting a full-sized buffalo head-on. Yes, it had been a total write off.

I won in court. In my defense, the Criminal Reports CR. Vol.39, page 404 –dealing with attempts to defraud by deceit. With the use of a past event from Dec. 10, 1962 in the annals of Canadian law in the British Columbia Court of Appeals. The sale of this vehicle was taken to our courts system, and this now fifty-year-old law was passed on to me by a British chap, a friend of mine who helped me win my case.

Law is based on the truth or lies, by the right hand. "So help me God." Here ye Here ye...He being a Lawyer himself had great humour in showing me how Canadian law works.

I still run into many relatives of mine in various parts of Canada and the United States. My generation traveled from Kamouraska,

QB, 323 miles, a 500 km car drive North Bay, ON. "Gold rush days" still bring the rule of registered claims for all by 12 noon, or you can loose it. These regulations were agreed ,then put in place namely to avoid past worst of case scenarios.

Canada's mining industry, and hunting-fish/fur trade is still a great part of everyday life with the brave souls that defend us in other parts of the world. The other side of the rivers or the lakes are the out-of-province regulation lines for us today. A well-known Algonquin Elder, William Commanda from Maniwaki, QC, whose uncle was well known Gabriel Commandant, a trapper and mining explorer of our region, was also an old army buddy of a fellow named Grey Owl, the book writer. The father of a child living with a lovely lady by the name Anahareo aka Gertrude Bernard lived in Temagami, ON. Her family lived just a short distance from our farm and acreage near Mattawa/North Bay, ON. Yes, a US movie that was made about their lives, as they were also both book writers who had traveled to Europe and had also lived in other parts of Canada.

The movie was largely about their conservation efforts just off of the Temagami -Mattawa Ontario river system. Anahareo was born in Mattawa, ON. They are both nicely shown at their museum, open to people on Voyageur Day, the powwow gathering for Algonquin's. Word does get around in small communities of our area, but it's well worth the reminder. Samuel de Champlain and his wife Helene Boulle are part of my lineage.

I was at an awesome powwow gathering 2006, in Mattawa, and got to meet family members and friends as we paddled our homemade birch bark canoe that day. Later, it was on display and my book's future publication was mentioned to their grandson. Elder William Commanda was there, along with the actress

Annie Gallipeau. They signed Makwa Kolts and my paddle next to the two stuffed beaver as a reminder of over-trapping in another movie called *Beaver People.*

North Bay, Ontario area has been the heart of many rendezvous and a home town of a few for many years.

As the father of New France, Samuel de Champlain had named the Algonquin's. He was introduced by one of his interpreter Jean Nicolet had learned the Algonquin language of this region. He also had married a nice Nipissing lady with whom he had one daughter, and she was part of the fur trade herself, till even now. She was married twice herself and then had a few children.

Jean Nicolet is publicly honoured by a plaque right by the "Gate of the North" in the city of North Bay.

A few others in Canada as well as in the US Genealogy, show that I and my relations have a lineage of twelve generations. I am still is part of this group, in which I have proudly participated over the years.

As a part-time job, I got to be part of the listed #10 North Bay Fur Trappers Association annual auction, being a helper in the sales end, hired on the spot by a Russian fur buyer. First thing for me was to fill a table with all the skunk furs as he was the only one that ever bought them, for a good price? Hunting and trapping in our back yard with my family uncles, I later trained as a meat cutter with Canadore College of Ontario, and I sure did put it to some good use in many provinces of Canada.

There was hunting, fishing, and prospecting in the Yukon, Northwest Territories and Nunavut while playing my role in the mining industry, namely feeding the hardworking crews from various parts of Canada. I met folks from our lineage who are still part of the mining industry, and fishing and hunting here and there, along with many Americans, some with dual citizenship, others just there for the hunt or to paddle rivers.

MÂS'KÉG MIKE's book presentation at local legion will have his family registry in place.

1901 census of Nipissing District # 92 Bonfield H1 page 18, family #144 Ouellete's father, who raised sixteen kids...I was named after him. There were quite a few of us with that name at Algonquin high school.

Michel Ouellette and Henriette Beauchamp m. 1858 St. Jerome QC – finally moved to Bonfield ON. First son *Michel/al born May, 13-1857 in St. Jerome Quebec, passed away Jan. twenty-four, 1921 Sudbury ON. married 1885 in Rockland ON. to *Elizabeth Beaudry born 1865/66 in Grenville Ontario, d.1997. She had at least two brothers...-Like many other family names, there were misspellings. This one was sometimes spelt Baudry or Vaudry. Specifics with accurate facts are listed online and in many books as well.

Behind cenotaph of 1922, War Monument Wall of North Bay ON. Along-side Branch #23 Legion. The Ouellette family name of Pvt. Victor J. Ouellette WWII, is proudly listed with many others of this area. Our uncle, Pvt. Hormidas was in WWI and is listed in war veteran's' found,-ancestry.ca search military record. A few of our relatives did join the military in the US. Published under the direction of the Algonquin Regiment Veteran's Association of Ontario, an Algonquin Regiment of 1939-1945 book called "WARPATH" by Major G.L. Cassidy , D.S.O.

Historic Society of Saint Boniface Manitoba provided me with a complete lineage book.

*Michel Ouellette and Elizabeth Beaudry/Vaudry m. Hull area in 1885. She passed away 1897, possibly at childbirth. His second marriage to Mathilda Diel was in 1904, in Matawatchan ON.

The stars mark Ouellette siblings in approximate order as we have relatives all over Canada and the United States. Of the six children there son *John Dudene Ouellette was in WW I. Oldest daughter called * Zolica married to John Doyle, *Raoul married to Laura Boulerice, *Valentine married to Damas Vaillancourt, *Marie Anne married to Midos Joseph Dollard *H/Artanse married to Philodore Picher, raised in Bonfield, Calvin and Mattawa ON.

A community and road was named after this family like many others had towns named after them for being there during a special time, or for being well liked, or because of their known bravery. The names were sometimes picked in a draw at school.

Many who remarried sometimes became relatives of our Canadian Aboriginals lineage or nationality. Their reasons for coming to Canada or the United States were due to war issues and their families sponsored them over to populate the country or just to work.

Michel Ouellette was born May 13, 1857 -passed away 1921, here are sisters and brothers.

*Honore /Henery Ouellette b.1861 and Matilda Picher m. 1884 Russel /Clarence Creek ON.*Leon Ouellette b. 1863 Russell County ON. Alice Gravel aka Lucie Morrel m. 1884 QC.

*Rose /Delimna Ouellette and Joseph Gagnon m. 1894 Bonfield ON.

*Philias/Felose Ouellette b. 1866 and Elaise - Haisse Piche/r m. 1885

*Azilda/t Ouellette b. 1865/70 and Isodore Picher Bonfield area m. 1888*Armidas Ouellette WWI b. 1871 d. 1920 Delima Rachel Doyles /Diel m.1910 Hanmer ON.

*Theodi/ule Ouellette b. 1872 and no marriage info, perhaps later moved to US

*Melchior J. Ouellette b.1873 and Clarida Picher m.1897 Bonfield St. Bernadette ON

*Emeli Ouellette b. 1874 and Adrien Lanthier m. 1893, Sainte Bernadette/Bonfield, ON.

*Alexandre Ouellette b.1879 Rose/Anne Cuillerier/Spooner, m.1907 Sudbury, ON

*Ovila Ouellette and Rose -Anna Rainville m.1907 Bonfield, ON

*Meloia Ouellette and Mary Larocque m. 1893 Bonfield ON

*Zoel Ouellette – lived in Bonfield ON. Wife unknown, perhaps medically challenged

*Victoria Ouellette b. 1885 et Napoleon Beaulieu, m.1901- Feb. St. Bernadette, Bonfield ON

*Clarida Ouellette d. 1881 and Joseph Achildes Vaillancourt m. 1872, then widowed 1881

This is my direct lineage of the Ouellette family tree, with some of the relatives involved with the exploring of the land for a safe upbringing in raising a large family grouping involved with farming. To various parts of Canada or course the US was easier to cross.

All aspects of trapping for that time are now still continued in North Bay, ON, which is one of the top ten areas for the fur trapping industry. It's still well liked by family members in both Canada and the warmer United States.

Passed on through generations for over 400 years as was done with groups all across Canada and the United States, were slight differences of laws, rules, and regulations to accommodate the area for the fur trade that was a main industry. Of course the industrial end of it picked up as time went on, to the high levels of today, with trap line acreage as our backyard, to firewood, mineral oil, and gas being main factors overall. The Government of Ontario put the Algonquin Park Act in place in 1893, Tom Thompson the Group of Seven to the public. Current part of the settlement allow Algonquin's to fish and a seasonal hunt of wild game. According to Stats Canada of 1613, the population at that time was about 10,000 and the Royal Proclamation of 1763 had reserved a large (unspecific) area of Canada for Aboriginals. This was out of fairness to those who had not been dealt with yet.

Then the government offered an open space to a large group of new Europeans. Crown land was not for personal use like golf courses or a backyard hunting facility, because of the relationship of cash and the buddy system of political influence. Canada does have plenty of space, and property is being distributed to this day.

Trap lines and mining claims were still part of life, passed on to one another and has continued till now.

My relations for many generations were a large part of our population for many years.

In this book there are pictures of both my great-grandfathers; the one dressed as a Chief is #4 on my lineage below. On his right side is the security guard Adelard Basil Leblond, husband of Henriette Sureau dit Blondin.

Family trees starts here, dit Auclair means clear water in French in the -A-B-C order

My dad Percy Ouellette was the son of (A) Henri dit Winskill – his wife was of the St Pierre lineage (D).

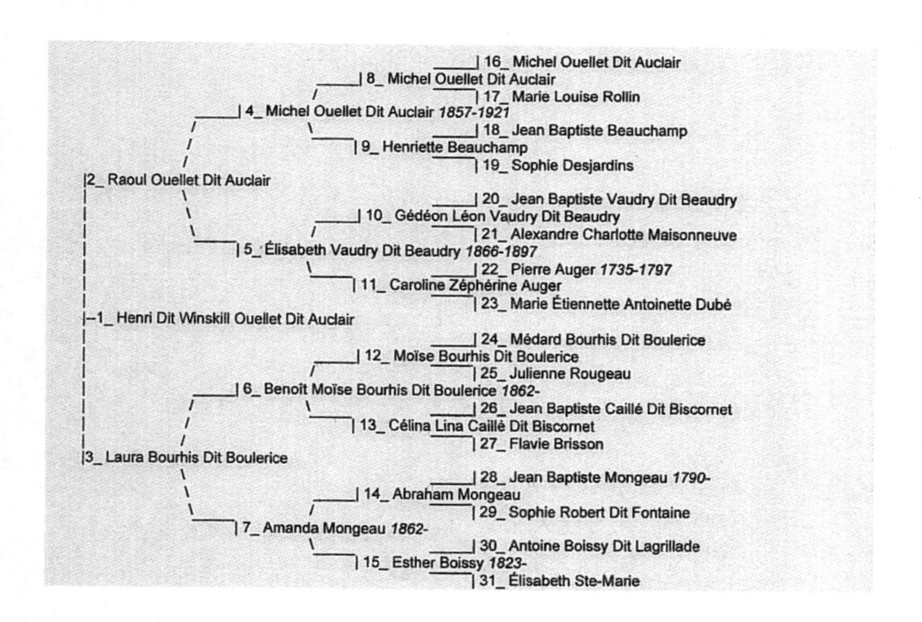

Riviere Ouelle is named after our family as they were heroes, there to protect an invasion.

With Aboriginals of the region, I am related to three-quarters of the family who are part of it.

"Hero's de La Riviere Ouelle" saving themselves, Catholic priest with a group of 40, fathers with sons on the list of the many hero's for that day. My first ancestors to Canada had different family names, this information comes with the compliments of the Historical Society of Saint Boniface, Manitoba. Algonquin - Mi'kmaq -Cree- Abenaki are my Aboriginal lineage.

Ouellette's lineage, town, and region found with our Catholic church's-library's help along with a few others' help along the way.

(Stars mark the six kids from the first marriage to my great-grand-father Michel Ouellet and *Elizabeth Beaudry)

#2 marriage Mathilda Diel b.1872 m.1904/07/07-Mattawatchan, ON. She was thirty so there may be kids. Raising kids on his own at aged forty, Michel's second marriage was to the sister of John Diel – They were raised in the Fort Coulonge, QC area, just across the Ottawa River near Pembroke, ON, where many met at a gathering place.

*Zolica b.1886 Bonfield ON. m.1904 - John Diel lived in Mattawa, ON and was, from Fort Coulonge, QC. Michel moved to Mattawa as horseback/canoes were still in use more in those days.

*John/Dudune. b.1888 (d.1918 WW1) registry of today's CA F 22 wing, North Bay, ON. Canada/US

*Paul Raoul b.1890 Laura Boulerice m. 1916- Chisholm Ontario right by Algonquin Park ON

*Artance Arthemise b.1892 m. 1914 Philodore Picher /Ouellet later moved to the Sudbury ON area.

*Vantine Caroline b. Oct.26 -1893 Bonfield, ON. d. 1966 m.1909 Damas Vaillancourt in Astorville ON.

*Marry Anne b.1896 m. 1916 Joseph Dollard Dugas. They later moved to Sarnia ON Canada

(**B**)Michel Ouellet m. Marie Louise Rollin Feb 1834 Mirabel (St -Scholastique) Deux Montagnes

Above's father list #2 Gabriel married to Josephte Biroleau 1785, in St. Eustache, Quebec

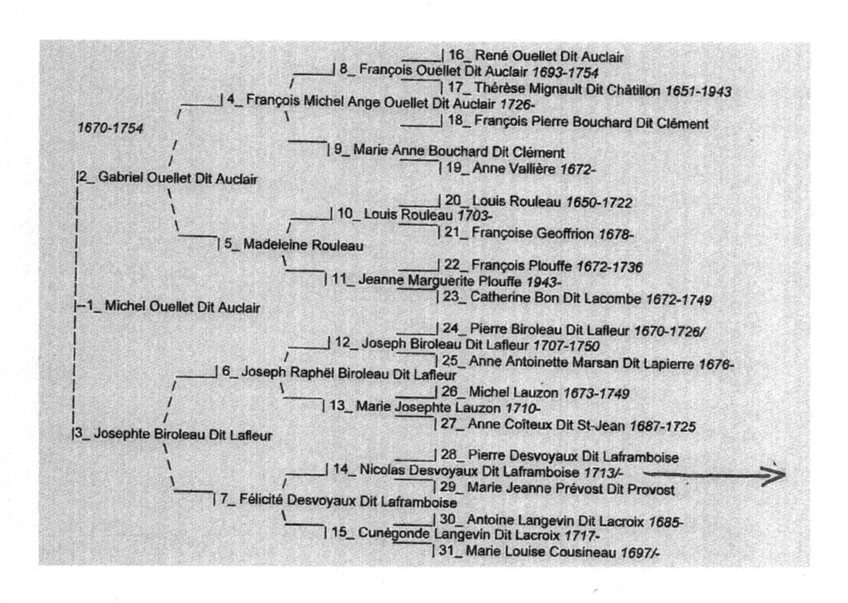

List (C) The above Nicolas Desvoyaux dit Laframboise was married to Marie Jeanne Prevost -----

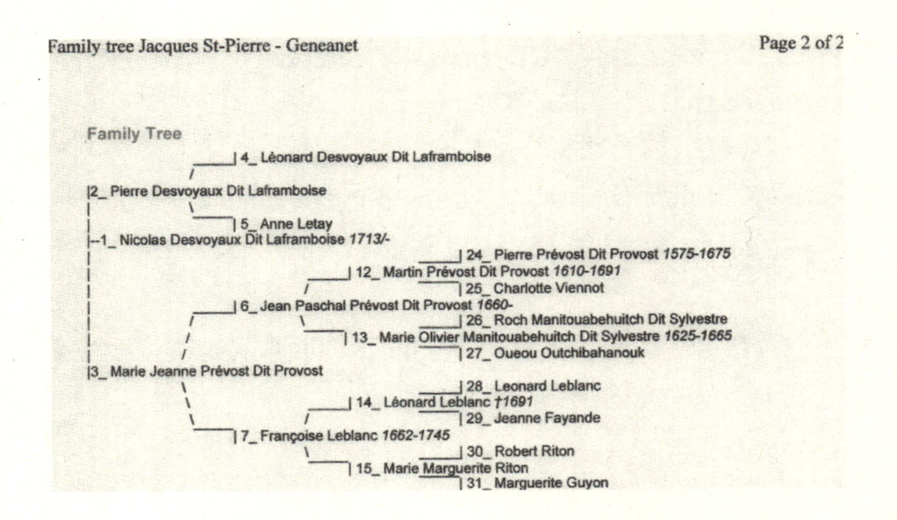

Family tree Jacques St-Pierre - Geneanet Page 2 of 2

Family Tree

4_ Léonard Desvoyaux Dit Laframboise
2_ Pierre Desvoyaux Dit Laframboise
5_ Anne Letay
1_ Nicolas Desvoyaux Dit Laframboise *1713/-*
24_ Pierre Prévost Dit Provost *1575-1675*
12_ Martin Prévost Dit Provost *1610-1691*
25_ Charlotte Viennot
6_ Jean Paschal Prévost Dit Provost *1660-*
26_ Roch Manitouabehuitch Dit Sylvestre
13_ Marie Olivier Manitouabehuitch Dit Sylvestre *1625-1665*
27_ Oueou Outchibahanouk
3_ Marie Jeanne Prévost Dit Provost
28_ Leonard Leblanc
14_ Léonard Leblanc *†1691*
29_ Jeanne Fayande
7_ Françoise Leblanc *1662-1745*
30_ Robert Riton
15_ Marie Marguerite Riton
31_ Marguerite Guyon

Great thanks to the following

Jacques St-Pierre's Family history GeneaNet.com online, my written version for those of PC inability.

Church of Jesus Christ of Latter Day Saints has a free online site and is easily verifiable. familysearch.org

Dedication to the 600 or so, here's a small list of the folks without whom I would not be here to write this book. Family name of one man of France and two lovely wives dedicated the upbringing in a tougher times back then. Said to be the Pioneers of "Riviere Ouelle," living on Iles D'Orleans also Beaupre. - La Pocatiere QC. d. 1721.

Both sides of my family; my father a Ouellette married to a St. Pierre. Sisters and brothers all passed away.

My mother's family name is Foisy and her mother was an Amyotte, direct relations from the very start.

Verifiable descent from the small, productive group of the first French folks to Canada as they intertwined to this day. They farmed, fished, hunted, and trapped along with helping one another as they lived with Métis/Aboriginals nearby. They are all part of building roads and the railway right across Canada.

The French and English both being part of the Indo-European language for many years to this day worldwide. Rene Hero Ouellet * and #1 Anne Rivet –#2 Therese Migneault, relatives across Canada /U.S. spelt in many French ways. Nicolas Lebel and Therese Mignealt /Mignot m. 1651, first husband they had four children and a total of fifteen kids together.

List #4(D) here is my father's mom, Melina St-Pierre of the farming community of Bonfield, ON

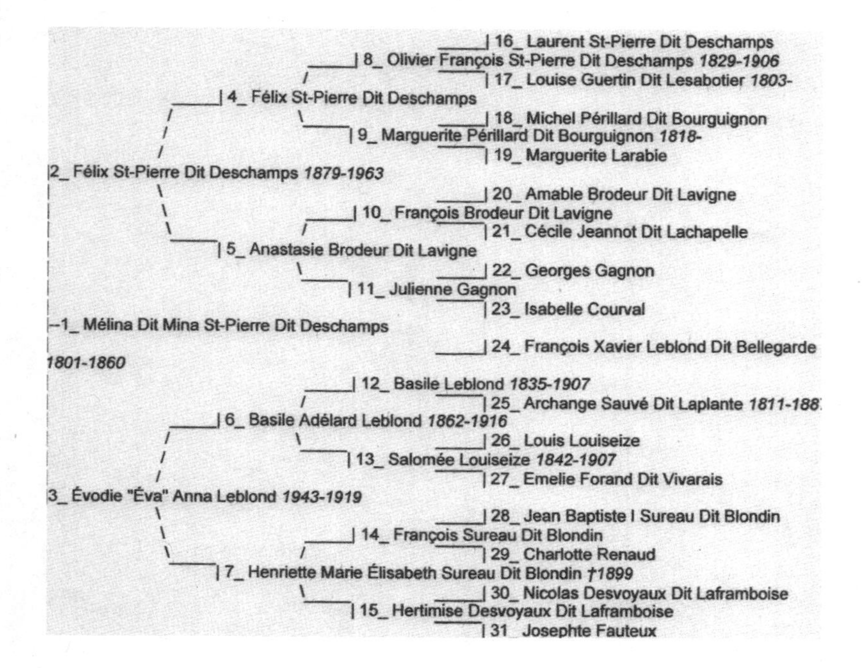

List #5 (E) the following is Laurent St. Pierre and Louise Guertin, parents of the above

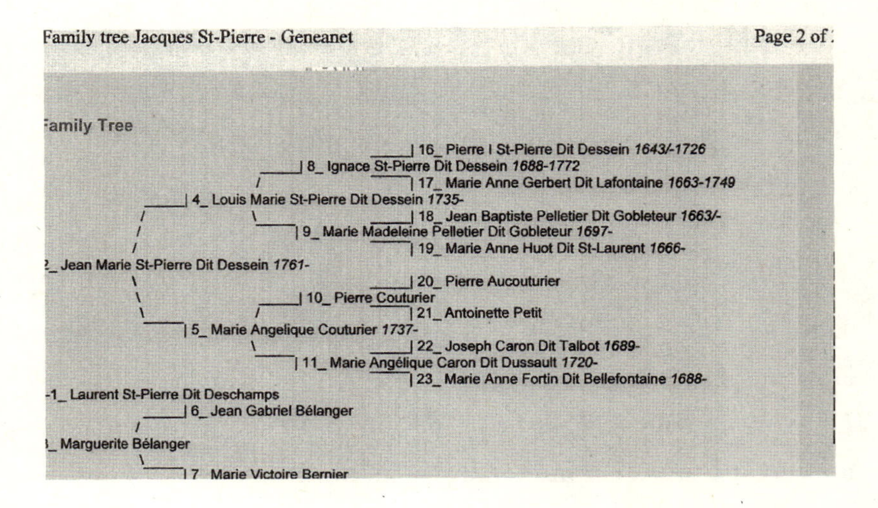

Family tree Jacques St-Pierre - Geneanet   Page 2 of

Family Tree

16_ Pierre I St-Pierre Dit Dessein *1643/-1726*
8_ Ignace St-Pierre Dit Dessein *1688-1772*
17_ Marie Anne Gerbert Dit Lafontaine *1663-1749*
4_ Louis Marie St-Pierre Dit Dessein *1735-*
18_ Jean Baptiste Pelletier Dit Gobleteur *1663/-*
9_ Marie Madeleine Pelletier Dit Gobleteur *1697-*
19_ Marie Anne Huot Dit St-Laurent *1666-*
2_ Jean Marie St-Pierre Dit Dessein *1761-*
20_ Pierre Aucouturier
10_ Pierre Couturier
21_ Antoinette Petit
5_ Marie Angelique Couturier *1737-*
22_ Joseph Caron Dit Talbot *1689-*
11_ Marie Angélique Caron Dit Dussault *1720-*
23_ Marie Anne Fortin Dit Bellefontaine *1688-*
-1_ Laurent St-Pierre Dit Deschamps
6_ Jean Gabriel Bélanger
3_ Marguerite Bélanger
7_ Marie Victoire Bernier

List #6 (F) Here are my mother's parents, starting with her father, Desire Foisy and Jeanne Amyotte

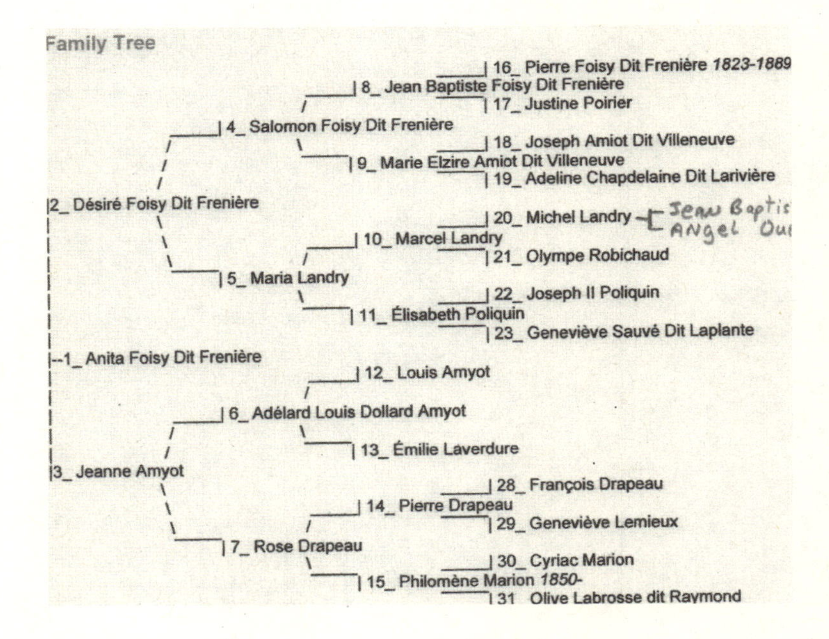

Family Tree

16_ Pierre Foisy Dit Frenière *1823-1889*
8_ Jean Baptiste Foisy Dit Frenière
17_ Justine Poirier
4_ Salomon Foisy Dit Frenière
18_ Joseph Amiot Dit Villeneuve
9_ Marie Elzire Amiot Dit Villeneuve
19_ Adeline Chapdelaine Dit Larivière
2_ Désiré Foisy Dit Frenière
20_ Michel Landry – Jean Baptis Angel Oui
10_ Marcel Landry
21_ Olympe Robichaud
5_ Maria Landry
22_ Joseph II Poliquin
11_ Élisabeth Poliquin
23_ Geneviève Sauvé Dit Laplante
--1_ Anita Foisy Dit Frenière
12_ Louis Amyot
6_ Adélard Louis Dollard Amyot
13_ Émilie Laverdure
3_ Jeanne Amyot
28_ François Drapeau
14_ Pierre Drapeau
29_ Geneviève Lemieux
7_ Rose Drapeau
30_ Cyriac Marion
15_ Philomène Marion *1850-*
31 Olive Labrosse dit Raymond

List #7(G) Here is Pierre Foisy Jr. #18 married to Marceline Emilia Chaput and follow up generations

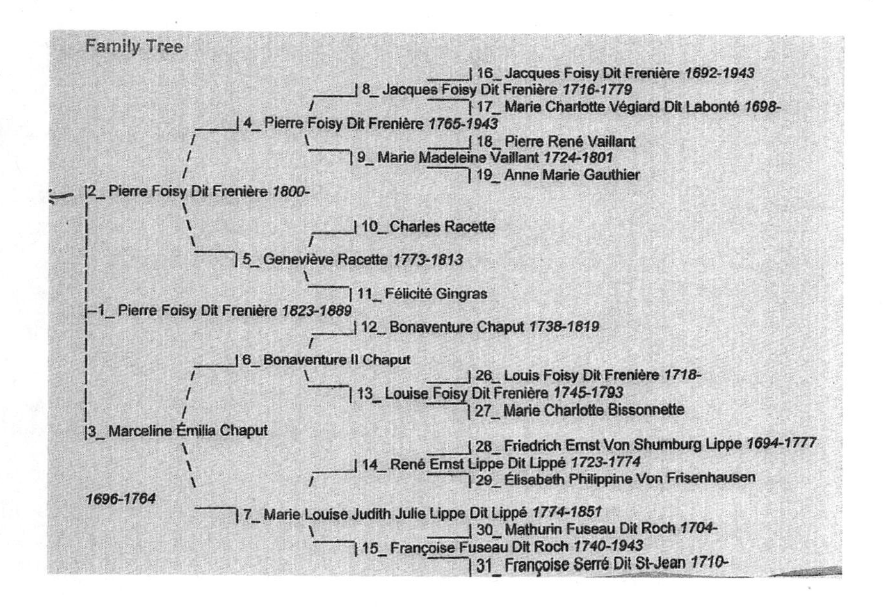

List #8(H) here is my mother's mom, Jeanne Amyot; family from La Passe, ON

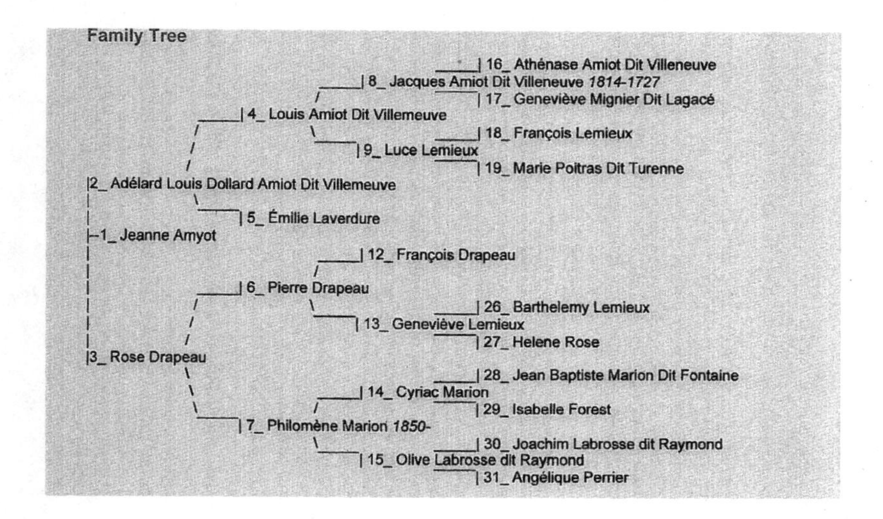

List # 9(I) to the completion of her Amyot lineage from Jacques Amyot 1814 is the above #8

My granny's side of my mother's family. With twelve kids, she is from this neck of the woods. Lineage #8

Our Lady of Mount Carmel church in Lapasse, Ontario, where Father Charles Arthur Ouellette served for seven years in the 1850s. He, like many others, crossed the Ottawa River, and he brought folks to marriage, baptized their young, and, of course, served at burials for family on both the Ontario and Quebec side. Made contact with midwives; since local women often gave birth with help from a Nanny at home, if there was not enough time to get to the hospital. Same applied to seniors on their last days.

Pioneer settler relatives of mine are still in the area to this day. They were highly respected as they represented a race of people of very strong faith. They built their homes in the forest, clearing their acreages of trees for the building of a house, furniture, and fire wood, with basic hand tools a horse and a bull. Back then, lengthy, burdensome trips still included the portages at river systems, over unfamiliar terrain to haul back the necessities to everyday life which they had to carry by hand, horse, and wagon / sleigh. Now and again they brought back medical men to aid the sick. My grandmother rode on a sleigh o a springtime morning. It was pulled by a horse and bull, carrying the basics to head to a new piece of land with a barrel full of supplies, and of course a rifle for wild game.

There was little to no chance for education – sometimes the youngest was able to make it to high school and on.

Many of the ladies married at a very young age to start a family, and raise their children. They fed the family and friends who helped build the house, but started in tents with outdoor washroom facilities. After the house, they then went onto the barn, for the raising of animals.

The outhouse for many was a dug-up area with shack over top that is still used to this day.

Wood-covered holes dug in the ground near the house, provided a colder version in the winter after the fall harvest. Dogs or cats were kept nearby or sometimes there was a young wild game critter as a pet.

Pork meat was put in a salt matter unit for storage, smoking, or storing in a cool place just before the cold would set in as a natural freezer unit. There was also jar pickling or canning in play.

Gardening was part of survival, along with trapping, which was done by many for food, not just for selling the furs.

Still popular in our family is the sea-pie/cipaille, a layered dough/bannock, with spices, meat, and vegetables in between. It is baked in an oven or buried in a cast iron pot/lid, often layered with dough over meat poked with holes, and broth/ water at the end. Then it's stuffed in a dug-out pit with plenty of hot coals underneath and straight branches top to bottom, which are covered with sand completely till the next day, with a pot of beans right next to it. Sea-pie and beans made with three choices of meat you have.

To this day, that same dish is served at the Bonfield church of Ontario as part of their summer event.

Close to one another near the "Golden Lake Ontario Indian Reserve," the "Algonquin Pikwàkanagàn First Nation" is just twenty-five miles south of the Pembroke area off the Ottawa River system that is part of today's Trans-Canada Hwy. As the roadways in existence today were the early paths off the rivers from A to B, where my dear grandma on the Amyote's side and many others were born in La Passe, Ontario across from Fort Coulonge, Quebec, just the other side.

Just across the Ottawa River at the now well-known military base area of Petawawa/Pembroke ON. A short distance away, is the Algonquin Park area of a well- canoed rapids, in a tributary of the Ottawa River system. The area itself is part of a cultural heritage, Petawawa's Algonquin meaning of! "Where one hears a noise like this." Gathering place. This gathering place that is at outer part of the Canadian Military Base area that has been in use since the early 1900's. Long-time use as a meeting place as well a narrow crossing area from the Quebec side both not far from one another. This being the

My grandmother's home town was at La Passe/Fort Coulonge, QC, near the well-known Algonquin trade site called, Island of the Matches "Allumettes Island" aka "One Eye" as this route was part of the transport route on to Ottawa, ON. Granny once told me about the summer job she had as human towpath pulled by a horse at a crossing, for a nickel a day. Yes, with her parents, and gear on a wagon/sleigh hauled by a bull and horse, I was told, all the way to Bonfield, ON for one of the many acreage offers. In around that time land was bought for twenty dollars, or traded for tree-cutting for government use, to help open up the roads and build bridges. Firewood haggling was part of the deal.

In the hundred acres or so, the anticipation of future gravel roads was part of the tree cutting for the ministry or clearing, with rules and regulations for their part of today's railway and road crossings in some places, along with gravel from one community to the next.

Sometimes, being the first to arrive or being a community hero conferred the worthiness of having your name used. The Mayor and staff's suggestion, or a school kids' vote was sometimes how a family name for at least a street or a lake was used sometimes they named part of another province.

List #10 (J) just below Anastasie D'abadie, an Abenakie granddaughter of Chief *Madockawando* Francois Xavier Robichaud, married to Julie Leblanc, was of the following Abenakie lineage.

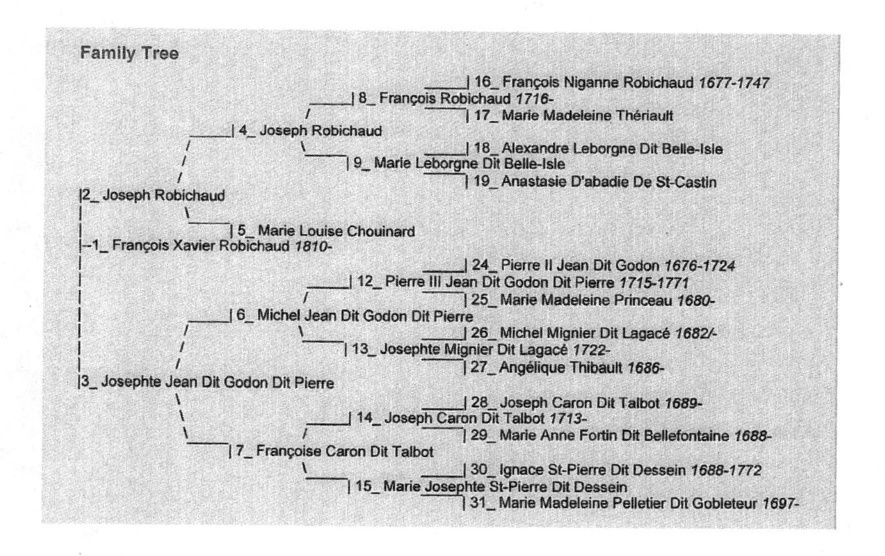

#10 On my mother's side of the upper page #6, Jean Baptist Landry married to Angel Ouellet. This is in relation to my lineage on all four sides of our family of U.S., Nova Scotia, New Brunswick, Quebec, and Ontario

Anastasie D'Abadie was the daughter of a Mathilda Pidwamiskawa and French Chief Jean D'Abadie. The Indians were brought under French influence by a man known as Castin in the New England *Chronicles,* who was a trader who did much to gain influence with the Abenaki, Penobscot, and other local tribes. He settled among them, and married a daughter of *Madockawando,* a Penobscot chief (Mathilda Pidwamiskawa)

Madockawando (born in Maine c. 1630; died 1698) was a sachem of the Penobscot Indians, an adopted son of *Assaminasqua,* whom he succeeded. He led his people against the English settlers in New England during King William's War.

La Piece de Resistance, DNA - Ontario Genealogical Society – Algonquin WWII vets registered Aboriginal Veterans Tribute Honour List online, WW II vets listed in Algonquin Regiment 1940 -1945

Historical Society of St. Boniface, Manitoba genealogy, North Bay library, Association of Ouellet-te's d' Ottawa. St. Pierre at GeneaNet.com online pride, free availability of Ancestry.com to participate and share with others.

On our way to PEI in 2009, my wife and I visited Elder William Commanda at his home in Maniwaki, QB. We got to see his years of achievement, his Wampum shell belt, his order of Canada medal, and the "Gold Key." We shook hands with a hug, both my wife and me.

As I proudly showed him my Algonquin card, with my WWII step-father's pictures, they sure looked alike. Of course, there was registered paperwork on his mother's name, since she was a Commanda of the Beaucage, Ontario area. She had married an Ojibway with the family name of Coucroche out at Whitefish Lake Reserve near Sudbury, ON. William smiled to say, you are my cousin, as we shook hands.

King Louis the XIV offered 150 lbs., to Aboriginal ladies to marry Frenchmen, to increase the Métis of the time. The men brought back young Aboriginal ladies to teach them the French language and religion, for their return to Canada. Safety and trust in numbers was part of life back then, as despite the treaty deals, there were still dangerous locations in trading. Jesuits did focus on ways for the Aboriginals, however they suppressed freedom for the safety in large numbers. Slavery existed for a time as the names were adjusted to suit marriage, eliminating the original in Canada and the United States. Back in the day, the rules and regulations of dealing with Royalty/First Nations in the fur trade were very strict, as the well-known explorers/Métis /Coureurs Des Bois/Shanty men/Runners of the Woods, all made it happen.

They married Aboriginals that were of my lineage from 1600s Algonquin, Ojibway, Huron, Mi'kmaq, Abanaki, and Cree. A Stats Canada census shows a population of 15,000 folks, mostly Aboriginals of the 1700s with whom they had made Canadian agreements. With a population mostly Aboriginal, treaty agreements were made in fairness for them and their future families of today. The sacred Wampum shell belts, are held in record of prophecies, history, treaties, and agreements to this day. With a lack of education, agreements were often by word of mouth, trusted, and then signed with a hand shake and a medal. These agreements are certainly open to adjustments today.

Back then till now the haggling mode, sometimes makes for less cost in a fair and trusted style to accommodate requirements. For many of the French people and many European groups, the greeting with a hug and a kiss amongst family friends both women and men is still done to this day, disliked by some a tap on the back or a firm hand shake is plenty good.

In our past, 400 years gone by, the Aboriginals of this land married the shipmen and soldiers to their king's daughters, for lives as farmers, and for medical care, church and school teachers, community leaders, and stores and supplies. Wagon builders provided survival care for all the brave women who gave us birth in this part of Canada and the United States. Political issues took place in fairness for the time, to bring them to what they are today. Waters were the number one element that brought the fur trade, lumber, minerals, oil, and gas to folks all over the world. Bilingualism has stuck with us as it did in the European and France section of the world.

Loving women in those early times some had rough days, raising many kids as best they could. They were encouraged to gain respect, as they intertwined with marriage to Aboriginals, for all their relatives and families to quietly carry on till this day. The Acadians and the Treaty rights of 1667, to the Aboriginal treaty rights of 1701 finally fell in place with agreements that had started earlier on. European fashions dictated dedication to the fur trade, along with the peace in writing of 1670, as the Hudson Bay Company offered "Trade Rights" to all draining river systems named, and placed it all on paper. Till now, the final process of the lower Algonquin regions of Ontario is finally here after many discussion for over twenty-four0 years, which included land and waterways by name, in the sorting of minerals and gas back then, and it does so even now.

From Quebec to all of the Maritime Provinces, the inter-marriage of French and Aboriginal resulted in the large portion of the Métis population of today's Canada and United States.

Like the European's method of creating expansion with peace by marriage, so did the Aboriginal chiefs of Canada and the U.S. France being the richest and most populated European country of 1789, language has been the decisive factor for politicians and churches that would run the show through election process. Of course, like today, English is also well liked but bilingualism all through time, has worked.

One well-known ancestor of mine was a man from France by the name Jean Vincent D'Abadie, who married Sachem Chief Madakawando's daughter with the Algonquin name of Mathilda Pidiwasmmiskwe/ Pidianski Nicoskwe (m. 1680 in St Castin QC). Jean then became the Chief himself to legally sign papers on behalf of the "Five Principal Nations" of the Acadia region.

Aboriginals did ceremonial group marriages. They took the Oath of Alliance, involving many aspects of the Fur trade of the early 1700s. Marie Leborgne, daughter of fur traders, was married to Francois Robichaud Jr. His father was a New Brunswick Premier.

Bilingualism of French and English has been part of an agreement in Canada and the U.S., as well for over 300 years. The two nations were settled over time, largely for fur trade reasons, which had a big part in creating today's border lines. There were agreements over the waterways, fishing, hunting, and trapping zones, along with lumber and minerals. North Bay, ON near my home town, is still on the top-ten list, for fur trade to this day, and its furs are still in demand all over.

Women and men at arms were part of these aspects that was passed on to them from the start in France and England where the middle class, like the voters of today made decisions.

Many of my cousins do not speak French or do very little as they are out west where even the poor Quebecois themselves struggled to blend in for the out-of-province pride of the work area, using their skills, which were in big demand. They also had the opportunity to learn other trades as they raise their families.

Aboriginals of Canada shared an honour with the French / English, along with many other ethnicities and lifestyles, just like the Algonquin members did in the WWII book called *Warpath.* They were properly registered in pride (as the Algonquin's of North Bay/Timmins, ON region,) just like all the other soldiers who defended our country. Canada's population was twelve million, with more than half of them women and children. Of the brave one million men who fought, many were of Aboriginal ancestry right across the country, as were the home front women with youngsters by their sides.

As a proud Legion member I received a document from the prominent historians of the Legionmagazine.com about brave men and women of the 1812 War, from a smaller population of less than 80,000 Canadians who kept our names in place. Métis' historical use of the sash along with the flag, is our ancestral reminder. Calling on all Canadians where ever they might be, we are reminded by the 116,000 WWII veterans to stop for two minutes for the wave of silence on the eleventh hour of the eleventh day of the year's eleventh month. Even now....

My dad's mom signed her name with an X, did not speak too much English and smiled a lot. She sure was a good listener. She knew about this when I was a kid as we had vets from our generations who had brought us where we are today. They'd been farmers for many generations, but education saved you from a nickname or gave you a reputation as a known one. From the Quebec region, five of our family's generations were raised in Ontario to work. The Ouellettes are across Canada/US.

At the many banks I have used in various parts in Canada, they always told me that there were twice as many in the United States but that here were 3,500 Mikey – Michel – le Mike – Michael – Mitchel and the slightly different Ouellette-Willet spelling. If you did know how to spell or write, chances are the English version of your name might have occurred at the Canada/US border.

CHAPTER 4

THE BIG CITY OF TORONTO

Each of us has his or her own story to tell. What drove me to embark in my adventures was my free spirit, with a curious appetite for the unknown. The bonus of being young, free, and single made me able to follow where my heart led me. At the age of fifteen, I was convinced there was something more to life than just the simple country lifestyle on the Hwy 17 Trans-Canada of Bonfield, ON. - Where I grew up.

So what more can you expect from a farm boy on his first trip to the big city but excitement and eagerness to face the world yonder. Of course, with my parents' permission in 1973, I got to leave home with my mother's older brother Viny. He was part of my baptism at our Catholic church to become my loving Godfather. Redefine pessimistic value to City life is just a wee bit different from country-style living: that is where I learned the new definitions, one of them being the morning dew. On this day I was visiting down-town Toronto, picking up groceries for an evening meal. An older, well-dressed lady walked past my uncle's side with a smile. I thought this must be a famous actress or singer. Arriving by the car, my uncle asked me to give him the book that was on the vehicle's dash board. So I just grabbed it to have a quick look to see what it was titled, and then all of sudden it hit my frame of thought who this was. It was "The Happy Hooker" written by Xaviera Hollander, who was now living in the big city of Toronto herself. She nicely autographed the book along with both our hands and a newspaper, then with a big smile, she walked away. "Don't tell your aunt" said my uncle she won't like it and either would many other good ladies care for what, just happened. "There was no worries as I gave him an okay.

Along with this Famous prostitute of the big city, there were many other famous folks near the Maple Leaf Garden arena: well, known hockey players and musicians. There were events and activities all year round. Yes indeed, the big city of Toronto gave me my democratic rights as an individual, and more oomph...

Toronto, yes, the big city of over half a million sure had a lot of new things for me to see. Of course, I met some nice folks from all over the world. During 1973's summer they had Yonge Street closed down for public walking and tourism. It was loaded with musicians and knickknacks of all kinds. Folks from all over the world were friendly to say the least, with the subway and buses for transportation. I later learned to drive on the 401 before my sixteenth birthday. Some sections of town had various types of industry, along with more bars than you can shake a stick at. Of course, many ladies that were very friendly towards me and my uncle would joke around saying they were not free. Relatives of mine lived in three sections of TO. My cousin took me out on a date with one of her schoolmates, to an indoor roller skating rink for my first attempt at balance on wheels for a few hours. The gals' tour included the many great restaurants of the area and nearby bars. As part of the tour there was the Gerard jailhouse, followed by the gay section of town, just off of Jarvis Street. Got to see some great sport events with a few bands at the Maple Leaf Garden in downtown TO.

I was in Toronto for about ten months. I learned to drive on the 401 Hwy and worked while enjoying life without a care in the world. Queued up while driving around from one end to the other; the rush hour sure taught everyone how to live fully with a high stress level. Though I was born and raised bilingual, the sign language format of Toronto's heavy traffic on the road and at the lights had a whole new meaning to me. Yes, I did save a few bucks and was looking forward to heading back home for the two-week Christmas family get-together for that year. It was a great winter-evening drive, sitting in a brand new Dodge van, right in between the two front passengers seats, on a half-empty, plastic, five-gallon pail with a cloth cushion, enjoying the view and listening to the radio just thirty miles away from our hometown, Bonfield. ON. With a stopover in North Bay to visit family and some of their friends. Happy thoughts went through my mind, for the season's greetings. Overall, I spent time with family, had some great food, played card games and did a little singing. It was soon to be a Christmas tale to tell that I would never forget, dear God.

It was one of those odd-temperature, winter evenings, changing from a nice warmer climate departure in Toronto, to a mid-afternoon after a five-hour drive that quickly got cooler that evening on our drive up north.

As the 200 miles drive home was nearly over, it was all of a sudden a big surprise to see unusually heavy snowfall. With evening darkness in place, minimizing the road's visibility just ahead of us, along with very extreme, slippery hillside moments, right there on the spot. All of a sudden, a large tractor-trailer in the opposite lane was coming our way, slightly swerving its back end on the curve of the road going past the yellow line. Our buddy was slowing down and, lessening speed as we were sliding side to side on the yellow line ourselves. We were heading down the slippery part of this slightly hilly curve, hoping to get back over safely to the good right side. With a look of defiance upon our faces, in regards to the outcome of a sure-to-run-into-one-another-hit, we all yelled, "Hang on for this, if we crash!" In a slow motion manner, with it taking about ten to fifteen seconds to impact, the drivers' side headlight of each vehicle collided head-on at about forty mph. Sturdy hands on both sides of my seat, I hung on tightly, sitting on my plastic five-gallon pail readying for impact. All of a sudden, with a big bang and a crunch noise, I was sliding backwards on the right sideways from impact with a simultaneous buckling of the metal floor. The back doors were forced to pop open.

A cut that would need nine stitches on the side of my eye, was delivered by the rear speaker, from the impact when the floor had buckled up eight inches to pop both doors wide open. I flew out somewhere out on the right side of the road, in a soft, thick snow bank with the pail along my side. I was just a bit dizzy when I awoke, virtually unhurt or so I thought, regaining balance as I staggered up the snow bank to the side of the road, just thankful to be alive.

All of a sudden, an OPP pulled right up close to me, yelling through the window, "Hey son! It is too hazardous to hitchhike this evening, so hop in the car."

Alright, so I did, then sat down at the front with breath of air coming back as I sat. He asked me what in tarnation I was doing out there. I felt that the right side of my head was sore and bleeding, and I held my hand out to show him the blood, saying we were just in a crash with a tractor trailer not too long ago. He pulled out his first aid kit to patch me up as he asked me where the van was after this impact.

I told him I just didn't know, as the radio confirmation that an accident had occurred was called in by the trucker who was right behind us, a quarter mile away headed to Toronto.

The tractor trailer driver had stopped up the road a ways as he flashed his emergency lights for the OPP to see in this brutal, evening storm. During radio communication he was told that another officer was on the way and he proceeded in search of the van for a more serious accident impact issue. Then, to complete the mystery, during the radio communication, he said that the van had run into the left side of the bumper of his vehicle with hardly any damage to his. Keeping in contact with the office, we then slowly drove about a quarter mile the other away before we finally saw the van I had been in on the opposite side of the road. It had slid and rolled over to the tree line with two fellows making their way back to the road, waving at us. Good God there they were, as we pulled to the side of the road with all of the flashing lights on to slow down oncoming traffic as the guys made it across to hop in the back of the OPP car. The officer asked how they felt and said he would take us all to the hospital in a short bit of time. Thanking the officer for getting us there, we helped him complete his report for that crash.

As I was in the emergency section of the hospital in North Bay, I was not the only one in need of medical care and stitches that day. My co-worker Tommy, who had been the driver was told of a wrist injury and got a cast. My uncle had a shoulder neckband holder, and an underarm pressure bandage, off of his right side. As we shook hands, we said, we'll see you later and I showed my stitches to them.

I got a ride home with my loving parents who had been called and came over to pick me up. My father had slid past and through the stop sign coming in. He had collided with an ambulance driver for a small bumper nick. For that Christmas season we had a good, thankful story to tell with a good feed. We sure were glad to see other family members that visited us out in the country at this time of the year. Always memorable.

We had a lot of parking space, a big kitchen table to eat at and play cards, and an extra room for sleepovers if need be. My dear mother's parents also had a place just across the road for needed back up.

Yes I had made a few dollars to play a game called Thirty-one; simply a combo of suits with the highest number winning after a knock on the table for the pot of fifteen cents each, or more if agreed at the start, with six or however many other players. We got three turns a piece, plus a free one. There was singing of Christmas songs, opening gifts (with hugs of course), and walks outdoors in the countryside after a big meal or a ski-do ride, and over to the next house to visit and say hello to folks we grew up with. Family we did not see too often was sure worth every moment.

Well now it was the new year of 1974 and officially I was sixteen years of age. This does make a big difference for many rules and regulations that apply to legal aspects namely in the workplaces and so on. Thanks to my loving father who had a spare set of work gloves and safety toe boots my size, I was able to be a labourer by his side. Body stretches at the days start before lifting concrete blocks, and mixing cement to then wheel barrel it to the needed scaffolding locations. Sometimes, along with use of the builder's pouch for the handling of nails pliers and a measuring tape, I'd reflect on the future potential of education, having gotten the ten, I should have gone for the grade twelve, as it is asked for nowadays. As here as a labourer and being a good listener, I safely learned many things in relation to the building industry. Oh well you've got to like what you're doing, so work gloves with safety toe boots with a lunch bag got me by with some of my loving father's friends' stories passed on at lunch. Wheel-barreling and handling the shovels had openings thought it was often just part time, as with timing you're not the only one out there, however, if you are good you're in big demand. Although sometimes boredom sank in and persuaded my restless character to be on the go, I still prayed to our Lord for my trailblazing adventures. My father reminded me that using my uncle's bull worker exerciser device would sure pay off, and it did.

CHAPTER 5
TOBACCO PICKING

1974

Having had a good taste of Toronto, the big city was a bit much for me at the time. Heck, was not looking to go back without a vehicle of sorts, because transportation was a bonus thing for work down in the big city. I told my uncle when the set of wheels comes, "See you on the road." Upon his family visit to North Bay, we got a chance to catch up on the latest talk about me driving here and there, and he told me that taxi driving in hauling folks around sure had a good dollar. With a vehicle in mind I had heard through the grapevine about picking tobacco leafs' potential. Physically demanding long day's work, pay was worth the while.

Seasonal work out in southern Ontario had openings every year. It was a busy time as favoured labourers sometimes from the states or other parts of the world came to put in time from the start to the finish. Population was a little low in the farming area, what with Mennonite families' horse and buggy living. They were still out there. Well, living at home with just a younger bro looking for his getaway from home time. Dad always had the work gloves for us with safety toe boots and hat. Having quit school himself, he would always say, "Are you looking to be a labourer for the rest of your life? Better do some push ups because you ain't getting free rent here." Percy, my dad, was friends with a few construction outfits and he hooked me up for minimum wage quicker than a bolt of lightning. Wheel-barreling cement seemed to always have openings. Lifting the concrete blocks to the chosen area from the big piles sure gave me the meaning of walking with a twist after a good day's hauling. Minimum wage at starting time was $4.25. Need I say more? Especially after earning that wage, a cold beer sure sounded good. My dad would say only one for you because you're under age, but you put in a good day's work. Laid off in the fall, I sure appreciated the chance.

As the first opportunity to walk into the Employment Service of Canada for a job application process was free and worth a try, it was nicely suggested by parents. On this day, with a big line up I was asked if I had ever picked tobacco. "No. Only as a cigarette smoker from stores," I said with a smile. Then I was asked if I had any physical defects or allergies. My reply was that I'd just finished hauling concrete as a labourer. Right on. Later I was told of this job opportunity: They trained you right on the spot with a free bus ride to the Delhi employment center of Ontario. It was heading for the tobacco belt tomorrow morning, so pack some clothes just in case you get hired. Some of the southern region farmers grew many other things other than just tobacco. Yes, we were told this job offer was very physically demanding and filled with many openings mostly for men. It was a chance to travel to southern Ontario to meet new employers.

Southern Ontario is an excellent area for the growth of crops. For the planting and picking of fruit and vegetables, farming families of today are still looking for trainable individuals to give day's work for pay. Room and board is included, with access to the phone to the picker's family, and the deal included a drive to local grocery stores for personal items that you paid for from your own pocket, if you forgot rain gear. Yes, cold beer on Sunday if you were old enough to buy some.

This bus was full of young fellows and a few ladies headed out for a plant and pickers exam – or so we all thought. In Delhi, ON. During our seven hour ride with a break in between, there was excited talk about this chance to be bussed here for work opportunities. What all did they ask and make you do? Just wait till we get there eh!

We arrived at the heart of Ontario's tobacco belt at a field filled with tobacco plants that were halfway grown, and mechanical picking equipment. Then we were asked to come out of the bus and stretch to get ready for the pickers' exam. So we stood out there as they grouped six to prepare for the team ride to just walk up to the machinery for a tour. Then there was a safety tour, along with an explanation of how to use the equipment. Right or left handed chose the opposite side, and then smooth picking techniques were suggested to fill the big basket by your side. We were to grab the handle and sit on the metal seat that was adjustable to suit your body height and length as you sat down at a one-foot level off the ground. Along your side was a two-foot square by three-foot high light-tubed metal basket on one side. Another one was just behind you for the change-over at the stop-off at the end of the picking row, which was about 1000 feet in length. We were told that not all farms had the same type of equipment used on this day; sometimes picking by hand with a holder was also done should machinery be in need of repair. Most importantly good pickers were required to work throughout the week, before the early frost between March and November, no matter what the temperature. Training at other parts of the farm was also offered, to complete the picking season to best accommodate all the demands for tobacco's seasonal sales market.

I had the chance on three occasions to pick, along with the harvesting cure process of the six-foot plants, thanks to EI each time. They would look at you and say, "Here –you're a picker eh!" Some of the pickers were from other parts of the world as their skills were worth their flight to do so. I did get to meet a few nice folks from Europe and the U.S. but that has now changed over time. Anyhow, that day as I was bunched up with the batch of the next six pickers, my bus buddy and I stood side by side hoping to go to the same picking crew. However this came out though, just to get a chance for the workplace and hard-earned money was pretty decent. So as we waited for our turn, we listened to one of the pickers who had just rode before. He said, "Pick as many as you can, twisting your wrist alongside of each plant. This thing moves fast."

With thumbs up, and smiling and nodding in thanks, we were looking forward to the opportunity to be one of the hired pickers this day. "You're next up. Have you been watching?" he asked. "Grab a seat. This is the bottom row examination, as the seats get adjusted higher up if you get to be a picker."

The driver took off with the EI representative giving him the go-ahead. "Here we go. Hang on tight with your feet on the stand."

I twisted my hat in just a wee bit then started to use both hands to pick on the lower left side, with a look over at the other five's techniques to improve my own. Watched the fellow next to me as he stuck both hands out to slide them on the leaf's stem straight down, to then close his fist in the grab and hold mode, to then toss the leaves into the basket. Prickly new ambidextrous definition. Wow, that sure seemed to work so I copied this method and did so non-stop down to the end of the row to fill that basket of mine. Indeed it worked. Upon our arrival, the trainee/guards gave me the thumbs-up with a smile, while switching the basket to fill the other on the way back. When we got back I had many leaves in both of the cubical holders. My buddy and I were told to go stand on the far side of the field as three of the others were nicely pointed at and just told to get back on the bus. That was it for them.

With a dozen or so of us standing out there while this happened quickly, we cheered each other on as some of the farmers approached looking at us in an odd and bizarre way. Got to say goodbye to the fellow I'd met on the bus. "See you around, eh?" Off he went with a farmer.

They pointed at their chosen ones to fill out papers, and then get into the vehicle over yonder. "You have just got hired. See you in a short bit."

A European gentleman with his wife approached me to ask if I spoke English very well. I stated, "Yes sir," with a smile.

"Sign the papers at the desk." Then pointed for me to wait over by a green Ford truck for the ride. "We'll be there in a short bit of time – we need one more. Nice Dutch and English-speaking family of tobacco farmers about two miles away."

Phew. Off we went to a good-size farm. The accommodations were beds side-by-each, all in one bunkhouse with a bell inside for wake-up at 4:30 to 5:00 every morning. Then I was introduced to his daughter, who drove the tobacco-picking vehicle. The man told us that she was in charge of all of us and if anything should happen to let her know right away. At that time she would stop the machine to give us fellows a chance to organize ourselves or help each other if someone gets hurt. "Enjoy your day. Breakfast at six-bells, lunch is thirty minutes at twelve-noon, supper at 5:30 pm. Do not be late or we will find another who can do this job for this very decent wage, okay?"

"Alright," we all replied. "Sounds good."

"So get some rest, you'll need it. We are not picking till tomorrow; the machine was in for repair and a tune up."

Anne, the man's lovely wife told me this was a chance to meet the others I'd be working with. "The older fellow is from England. He is well experienced, so have a discussion with him about it all."

Standing by the harvester was a bearded man having a cigarette. He was dressed in an old, blue raincoat and was wearing dirty, green work pants, black rubber boots, and a bright yellow hat with the flap pointing up. This Englishman approached me with a bit of a swagger, hands in the pockets, and a big grin on his face. With a humorous British accent he said, "Allo allo allo. Good day mate." As we shook hands he said, "I'm John so and so from Manchester, England."

With a smile as we shook hands I replied in my best Brit accent," I say, old chap, only made in Canada."

He started to laugh. "Righto," he said, using an old British expression.

Well it sure worked. We were instant friends and continued the day with a few laughs and talk about our backgrounds, followed with do-and-don't like and dislike of machinery. We agreed on money being the reason we were all here. Pip pip and all that. John had his own car there with a few nearby friends in this neck of the woods as he had picked for more than one season. Surprise when I told him my first language was French. "Blimey!" he stated. "Well so are the other four pickers who hardly speak any English very well, who are apparently Quebecker. I told the boss he'd better find someone that speaks English so I have at least someone to speak to or Bob's your uncle."

We laughed in realization that to at least communicate during a long day's work or to have a cold beer together after our sweaty shift was okay.

The others pickers seemed to keep to themselves alone, overall. As he had picked for a few years himself, John knew his way around and showed me a few tricks. He told me what to watch out for to safely pick without hurting myself. He also showed from start to finish the area where we'd be handling some of our tobacco, and took me out for a drive with a fun tour right up to the Tilsonburg Hotel/Pub just around the corner. As he got there, I just told him I was too young to go in there.

"No worries. Go sit by the side entrance where I will sneak you a cold beer as you listen to the Canadian musician."

"Okay, right on. Sounds good." Not long after, as I sat at the side entranceway on a concrete stairway by the open door to listen to the music, it all of a sudden ended. Must have been break time because this man wearing a cowboy hat with a white T-shirt and a cold beer in one hand and a glass in the other, came out through that door. As I sat there telling him that it sure was good music to listen to, he poured half his beer in the glass with a "Cheers" as he gave it to me. He asked me what the hell was I doing out here. "My name is Stomping Tom Connors," he said with a laugh.

I told him Michael was mine and that I was only sixteen years of age and had gotten there on the bus to come pick tobacco. "My British friend said he's going to sneak one to me. Ha Ha."

He lit and smoked a cigarette, telling me he had also picked tobacco himself. We sat around for a bit and he told me he would next play the song next called, "My Back Still Hurts When I Hear Those Words." He finished his smoke and got to meet my friend John as John snuck to the door with a cold one for me and I offered him half.

"Yes, I did hear that song before," John said with a thumbs up. Shortly thereafter an introduction to the second set followed with song Tom promised. I listened for while then waved goodbye, because we all knew we had an early start the following day.

Off we went to the farm to get some rest, and I fell asleep with a smile that night.

The next day at five o'clock, the bell rang very loudly non-stop for two minutes. We got up and got dressed, getting ready for action till lunch as we entered the house for an oatmeal and toast breakfast. Stretched while outside, shortly thereafter, with a bit of pain in all parts of the body from the day before, mostly in both wrists. We were introduced to the daughter, Maggie, as she warmed up the picking machine, telling us she would start off slow to give us new pickers a chance to get the hang of it, and to put a productive, safe work day.

Here we go with just one seat left to choose from on the machine as I hopped on good and tight. Off we go to the next picking field. I was told by the other pickers that the leaves were wet so I should tightly wear the hat slightly bent over the left eye for protection from the tar sap as I picked. "Just pick as best you can for the first week, till your muscles adjust to this labouring work day."

Row by row of repetitious picking to fill the large baskets was on the go, lifting the basket onto the waiting trailer at the end of each row to bring the leaves' to the next process of smelting in a kiln for a day or two. At this site the cigarette leaves were quickly sown together with a fish line between two sticks on a three-foot row and sent up on a leather mat to be hung and cured in the kiln for a length of time. This was done by a trained group, mostly ladies.

Then after supper, we got an offer to haul out chain-style, one to the next, for five dollars and cold beer. The next step was sorting out by the best grade, and then off to the storage shed.

I was told that the best leaf piles were worth a dollar apiece at the time of the early '70s. With today's price of over ten dollars a package, the process scams are gone beyond reality for this silly addiction that is still allowed to poison the public. I sure enjoyed every one I smoked, but am now going on eighteen years without as of the year of 2014.

Hands were covered with a thick, sticky, black tar gum by lunchtime, mostly from the early leaf picking. It was very difficult to remove afterward. My coworkers said they used

Whatever was offered by the outdoor shower unit as our outdoor shower had a cold-water garden hose with a nozzle tied to it, covered by a scary, stained, dirty, black canvas tarp for a curtain. They had a pail of cleaner to help take some of the tar off your body parts, mostly your hands, with alcohol and a scrub brush. If you did get cut they recommended the use of a rubber glove to prevent infection. It was worth every dollar of time put in, with a nice one-day off a week, or an offer to do without and work, but only if you wanted to. Every day was a long hard work day. In my eight weeks there, two fellows had quit. Another of them had left after he got a half-inch cut on an eye from a leaf, and it got infected. Worse than a black eye per se. All in all, just to say this experience would last me forever, when I got an offer to pick some more, I did so later on. When I got home though, it sure was a nice break with an account high in cash to get a car or something. I took it easy at a shared apartment for a while, for my independence.

CHAPTER 6

TEA BAG ANNIE

With the winter of 1974 already past now, to the summer of 1975, I was bored out my tree. With not much work experience, or so I thought, and handling labour work well, I should have finished my grade twelve, eh? Toronto as a carpet layer's helper, wheel-barreling part-time, along with picking a little tobacco sure was physically demanding work. Some of my friends who had completed their education found that it had proven itself.

Filling out workplace job applications including a resume, a lack of education, I felt my file was tucked away. Other workers told me you might be eligible for unemployment insurance if you were seasonal with hours to quality for it. Wow, this was going to be my first time, but there was going to be a waiting time to verify that enough weeks were in, to keep me looking for work was part of it all, eh? So, I thought I'll just wait and see what happens while at my parents' place. I felt they would put up with me for a little while longer, and indeed paying rent made sense. Upon my return to Bonfield, my loving mother had reminded me once again, I should not have quit high school, as grade twelve makes a difference in the workplaces. My parents were charging me rent for living at their house, so look for a job okay?

Well this encouragement to keep busy looking for a job had always worked for her, and sure did so for me. Gosh golly gee Mom. Dad said, "He still has his gloves so I'll keep him busy to give it some effort to get out there looking for work. It is early summer, after all. They're always looking for labour. Don't worry, I'll keep my ears open." Physical labour sure had its benefits – lots of openings if you kept asking.

With no vehicle for transportation while out in the country, we were still allowed to hitch-hike as I grew up. This day, with a call to a nearby relative by my dear loving mother's suggestion, yes, I grabbed a ride to North Bay with a cousin who was on his way to town. Jumped out at a little park and sat on a bench by the branch #23 Legion looking at the well-respected military statue, wondering what to do next. Yes, it was sunny that day with birds chirping as I walked by to see the names of our honourable war vets on the monument of a soldier holding a rifle. I saw a Ouellette name on it, wow. Are we related? Oh well, from the stories we heard those men and women had given their lives for us to be thankful to all the vets with respect each Remembrance Day on Nov. 11.

My friend finally showed up as we sat on the park benches. "Well, we are only sixteen years of age did you hitch-hike here?"

"No, my aunt gave me a lift."

"Well, that's hitching a ride." We laughed and that is when my buddy suggested to do so again this time, all the way to the east coast of Canada, boy, Nova Scotia. In putting some thought to it, we sat talking about what it was like out there. As he was raised in that area, it was worth a look-see. Around a thousand miles away – sure would be a lot of hitchhiking, but it's worth a try.

I have an uncle out there and I thought this would sure be an adventure to remember. We could live in a tent in their back yard with their permission. I had read about Nova Scotia in school and seen it a few television news about the Halifax harbour right on the ocean. It was a chance to meet some new people. Sure. Why not? I had never experienced the other provinces and now I was going to see three others going through. Okay then. Sounds like a plan. We shook hands.

So I discussed it with my parents and they said, okay be careful. They told me we had an aunt there too and that I should make contact with her when I got there and gave the number. They did not approve of me hitch-hiking but we were three guys and the oldest was twenty years old. So in preparation, we packed a lunch, snacks, and water, and one pack-sack full of clothes. Off we went to be on the road again.

On that fateful day, we were ready to go and I announced to my dad that this was it, I was heading to the East Coast for a visit and maybe some work like picking potatoes.

"We'll be careful," he said. "And have a nice trip."

Mom did not approve too much of me being in the company of two long-haired hippies and all that, you know. "Be careful on those roads when you come back we'll have the work gloves all ready for you son."

"Sounds good, love you both." Hugs and a smile.

Both my friends were from the East Coast, so they said their uncle would most likely give us job and put us up for a bit. Plus I had an aunt to visit upon my arrival. With twenty dollars in my pocket and a loaf of bread, along with a jar of peanut butter, a chunk of baloney, and a can of beans, my tooth brush, and a blanket. Oh yeah, we were carrying our accommodations –we had a six-man tent that we'd take turns carrying.

Since there were three of us, thumbing together was not a good idea, but we did not give up and made a cardboard sign.

We got stuck for two days, thirty miles out of North Bay in the small community of Mattawa, Ontario right by the Ottawa River. This was going to be tougher than we thought. So we walked past the town limits and we would take turns standing with our thumbs out and the sign. This was my first time hitching for a ride for such a long distance. Well, night-time came and we put up the tent just off the side of the road and ate some of our sandwiches. This was a good routine for a rainy day emergency, just in case. We had a flashlight and we shot the breeze about what lay ahead, with better luck tomorrow.

Finally, the third day we got a short ride for twenty miles up the road – some nice folks thought we were local from a nearby community. It did not matter, at least we were on our way. We figured approximately 930 miles to go – technically just a two-day drive or somewhere there about. Then a next ride – nice people picked us up despite our appearance and smell. Hah aha. Fourth ride, eh.

Finally we got to Ottawa. It was nightfall and someone told us of a hostel at an old jail site. Wow, you should have seen this place. It was like being in a creepy horror movie with squeaking doors. The cells had ordinary doors but the locks had been rendered inoperable, with some doors squeaking loudly as you entered. Well, a good sleep for the next day when to the main highway we headed down, sign in hand with a good feed.

Got a ride right away in the back of the KFC truck – a little cool but at least we were moving on that morning. With the flashlight in hand, we then got dropped off somewhere near the Quebec border. We got rides in the back of a few trucks and a station wagon. Yes, it was very true, three of us sure did make it a lot more difficult than we ever expected, but off we go without giving up, all the way, with a few short rides, a little at a time. Nice folks laughed as they said, "All the way to Nova Scotia, eh?"

Upon our arrival at the local college area of Fredericton, we were nicely told to sleep at a nearby church. This was just great and we thanked everyone there as we prayed for a ride that morning to make to our destination, Halifax.

We got a little ride close to the highway, with directions, and made sure to have out a sign that was easily read.

Twenty-four rides into our trip, on the Trans-Canada just outside Fredericton, New Brunswick there she was, in all her splendour; a smoking, oil out of the back tailpipe truck with a kind of a forest-green homemade paint job, and one of the doors held by a rope. Written on its side painted in big white and brown letters, were the words Tea Bag Annie. Wee haaa! Wow what timing! we had only been out here but an hour.

As he pulled over we each saw a look in each other's eyes as if to ask if we were taking it. We all nodded yes.

We had full stomachs from the accommodations of New Brunswick – praise be to the Lord. Although it stunk a little more than we did, no worries with the windows down.

The kind soul who picked us up was a long-haired, East Indian man with a bit of an accent. He was picking up all of the hitch-hikers as he was on his way to Nova Scotia, for free –no charge. "Wow, what a lifesaver," we said with a smile.

"Where you guys going?"

Whoops our Halifax or Bust sign was still in the bag. Hahaha.

Good hand shake with a big thanks for the best 25th ride of them all. Phew.

"So are we going to make it?" we asked.

Thanking that fellow, we learned we were three of many he had picked up. Then we rode in the back for five hours.

Stretching out on our arrival with hands in the air we all yelled, "We finally made to the East Coast!" Now in Dartmouth, Nova Scotia, the City of Lakes, we were dropped off to a scenic view right by the Mic Mac Mall. With our time spent at the local mall to meet folks of this neck of the woods, we noticed just a wee bit of a difference right off the bat. A little over 1,000 miles in distance away, we were still in Canada. Yes, a bit of thick Canadian accent aka the Maine-ish type of the coastline with a Scottish touch. While we were at the great mall talking to a Newfoundlander that day, he told us the Newfoundland version was even thicker, bo'y. Told him we had hitchhiked all the way from Ontario and he told us, "You're not easy."

That nice afternoon we called my buddy's uncle, who then took our camping gear to his house where we got offered a parking space in the backyard at a cost. As their kids were all grown up and living on their own, we were given permission to knock on the door when it was time to eat and use the washroom. We kept quiet with no parties out back in exchange for meeting some of the nice, nearby Mari-timers. Without a vehicle we sure got to walk and take some of the trails where there were the apple orchards of the Annapolis Valley, for a great choice of tastes.

Deal haggling included mowing of the lawn, watering the nice garden, and washing the car. Best of all we got a tour of the coast then were offered a few rides out in the uncle's good-size boat. Then we helped him empty his nets and crab traps, loading up a few types of ocean critters for the feed that his loving wife would give us help. Occasionally, we helped pickle food in jars with her. Best of all, I got to cross the historic MacDonald bridge because the uncle had a contract with a few ships in the Halifax Harbour that came in to pick up the garbage for disposal at a local dump site. We stood there dressed in work gloves, rubber boots, and holding a half a broomstick with a sharply bent nail at the end. "What is this for"" we asked.

"Well," he said with a smile, "just in case the rats as big as cats come after you, just poke him in the head first. Okay?"

We were verbally given the rules and regulations on loading the bags left on the deck for us to toss in the truck and trailer along our side.

Well, I did get to see a few huge rats but with my stick in hand, they ran to another nearby pile of bags.

This was my first time going across such a huge bridge with an awesome view of the ocean waters that day as we stood on the dock with an awesome view of that fine-looking bridge. When we were just about finished, the sharply dressed captain of a military ship walked by us. He then asked us with a smile, "How old are you fellows?"

The two brothers spoke to him, telling him their age. "I will be seventeen in another month!"

He then stated, "You can join the military and do better than this," as he kept on walking by us. "Have a great day fellows."

Oh well, we had made a few bucks along the way, with an awesome fun visit with my buddies. Uncle took us to Peggy's Cove, NS, for a good taste of the well-known clam chowder as we toured around that rocky view of the St. Margaret's Bay of that east coast area. Man oh man. That day, as I viewed the bay while walking on the rocks, I was told to walk in a safe manner and to watch my ass for this was the quickest rising tide in the world, with the water level increasing in a very short bit of time. "So get the hell out of there and take your time walking on them there rocks."

This was an amazing first time and I was surprised in watching the high amount of ocean water waves collide with one another, as the tide rose in a short bit of time. So I made my way back to shore along the rocky edge. We then made our way back to Peggy's Lighthouse for a tour of the kitchen with a big smile. Of course we did get our bowl of the well- liked clam chowder with a tea biscuit. Uncle told us of the popularity of this fishing area but to be very watchful for the tide factor.

He and his wife, that welcoming couple, treated us very fine with East Coast hospitality for the three summer months of this adventure. Some of my well known ancestors were from the area, like Catherine Petitpas/Bugaret whose father had started the fur trade in then-Acadia back in the 1630s. Claude Petitpas Jr., her son, was an interpreter/court clerk who kept track of the wood sales along with furs being sent back to Europe. Like many other pioneers of the time, he was married to a Mi'kmaq lady of the area, as the tribal concept is in existence to this day in that coastal region. Encouraged to have large families in those days is why a large number of Canadians and Americans have relatives coast to coast, with a few now in various parts of the world.

Yes, I had gotten the opportunity to meet for the first time an aunt of mine, whose husband was in the military. She gave me the opportunity to call home to let my loving parents know I had made it in twenty-five rides. Only found that out later, to write this book.

CHAPTER 7

THE SOLDIER

Canadian Armed Force 1975/76

Me as a graduate from boot camp in Cornwallis, NS in July 1976 and was honourably released in October of the same year

At boot camp in CFB Cornwallis, NS. 1975-76

Well the captain sure was right, the opportunity to join the military at the age of seventeen was all true, as he had mentioned to me while I was cleaning up garbage at the Halifax Shipyard. His remark that day with the awesome view of the Halifax harbour that I could do better, this had surely encouraged me to join. As I confidently looked forward to return home and a visit to my familiar surroundings at my parents abode, the Canadian Air Force was #1 top of my list to check out upon my arrival back to North Bay. Enjoying the ride on a train back to town as I'd had never been on a train before, was sure worth every moment. There were spectacularly impressive views with great staff service. It sure beat the hitch-hiking aspect of the twenty-five rides, to say the least. Upon my arrival, my parents asked me how the train ride was. I told them about the view coming in, and how I'd met some nice folks as well, and learned a few things. They were glad to see me after I'd traveled such a long distance, and they cheered me on with a few big hugs. I told them stories of my time spent in the coastal region and how I got to meet our military cousins and my aunt out on the east coast, though her husband was away on a ship. I said that I had gotten to see some nice pictures and talked about the many things we got to do while in the Maritimers area. There had been slightly different food from the ocean, and I shared the humour of everyday life on the east coast of Canada. With no pictures in hand it got me to thinking I should have invested or borrowed a camera for a few pics.

Well now, here I had a surprise. They told me that my first letter had arrived from EI of Canada. Son of a gun, something about a lack of hours for collecting Unemployment Insurance at this time. So that was okay, I still had a few dollars in the bank from the tobacco field picking, and I sure was thinking about joining the army. In memory of being the best of my class for the Youth Appreciated Week of 1969 (November 10 to 16) I had received a hand shake from the War Vets Association in front of students in my grade five, with a certificate for my written poem and a bonus of a four-color pen, in encouragement to keep on writing. With heavy security, we were in a small group that was allowed to visit on this day. It was a student tour of the NORAD Underground complex at CFB North Bay, with a great view of this huge bunker, said to be 680 foot deep with a one-foot-thick steel door at the entrance way that opened up. Led by a veteran and a security guard, we walked through the laneway on an exemplary tour followed by a movie clip. The Algonquin Regiment color display was in place with, "Lest We Forget." This was the pride of the Canadian Army's 33rd brigade Primary Reserves of the Algonquin Regiment A – B company, nicknamed " Algoons, Gonk." Motto

"Ne -kah- ne -tah (we lead others follow) on the march. "We Lead Others Follow" was displayed on their flag.

I had titled my poem, "Remember This Day," as I'd been told by family members that we had war vets on both sides. Our relatives' pride in learning more was why I was here. That is why we wore the poppies on the eleventh hour of the eleventh day of November every year, for a moment of silence in reflection of prayer. It was a reminder of our WWI "The Northern Pioneer" vets' blood, which had been spilled like that of the many other troops who had put an end to hostilities with signing of a treaty. This time of peace was followed by WWII's total war involving folks from thirty countries sure opened our eyes. We our thanked everyone, especially our teachers, for this amazing tour to pass on the word to let others know that our freedom exists today because of all these war veterans – so treat them with nothing but kindness. Post-Traumatic Stress Disorder awareness, I would find out, was studied and part of my test after my training afterward.

I was looking forward to my seventeenth birthday in the year of 1975 in more ways than one: so that my application for the Boot Camp session was approved – firstly to fill out the forms, followed with a full medical right off to no problems. However, needing my parents' permission was part of the deal. Got the story ready of my adventurous thumbing time to the coast for those three months, where meeting the ship's captain had a military influence that paid me off to this day. Because of my service, my resume would always get me in the door to wherever I apply. My poem sure had paid off and I'd likely get a chance to finish my grade twelve.

That morning, I told my parents their signature was required for me to join the military please and thank you. "Are you crazy?" my dad asked. "What's the matter with you? That is not the best idea, son. What about your grade twelve? Joining the army is way different than what you have been doing, so are you ready? Well at least you are physically fit – with a nice talk to your aunty about do's and don'ts you may get your grade twelve out of the deal. So it is your life, whatever you decide it is all up to you."

The following day, I got a ride to town from our country household in Bonfield. Indeed education was important though my brother and sister had also quit school early themselves to head out to the workplace.

Well, here we go again as they gave me paper work to be signed by my parents along with a complete medical, then a test for work applicants to join the Canadian Armed Forces. These form were then to be taken to the Main Street head office in North Bay. For the waiting time of just a little more than a week. Here we go again, with a big hug to parents. I was asked to give them a call when I got there, and to behave on this training because this is sure different than what you've learned. I got an offer to take a French course in Quebec as I spoke French very well. However, I still had an accent and English was usually my choice of language.

Wouldn't you know it, they sent me right back to Nova Scotia, although this time to CFB of Cornwallis, right off the Bay of Fundy. Now this being my first plane ride ever, it was better than the train ride, or so I thought upon our departure. We shortly landed in Canada's capital at the Ottawa airport and I had a nauseous feeling of excitement and dizziness when the fellow sitting next to me with a smile handed me the puke bag, stating my face had turned green. The flight attendant gave me water and a pill just before the next take-off to the East Coast, and she told that it happens to others as well. The passenger smiled as he later offered me half his beer in a glass to relax me from the motion sickness, which was all new to me. The meaning of emulsification process was redefined for me that day.

My body had nicely calmed as we landed a few hours later, memorably glad for the first free ride I ever got, with a great view of Halifax from the air on a sunny day. Thanked everyone for this flight upon our arrival. They had a sign as they called out our names from the mixed-in other civilians' on the flight. What a marvellous view by plane of the mountainous range in this valley's growing regions. Tickets and baggage in one hand, we then hopped on a bus from Halifax on our way to Cornwallis for the Boot Camp trial with a few others in a story of the 5Ws for that day. As some of the crew were military children or had been in boy-girl scouts/ rangers they knew some of the things that would prepare us for the next few weeks.

Sure was all true about the importance of their disciplinary measure; fines for mistakes made, or kicked out if you messed up, or just wanted to go back home. Here we were on the bus for a two-hour drive to the Cornwallis base, feeling proud and looking forward in anticipation of what would take place next, when all of a sudden they pull the bus over in the small community of Sackville and ask us to step outside of the bus to then go alongside in one line up. With all of our baggage being taken off the bus, we looked at one another wondering what was going on. This appeared to be a military storage area where they kept and fuelled part of their bus service should a breakdown or other emergency issues occur. It was not far from the airport in Halifax. A sergeant with a dog walked towards us on this October evening, and asked us two simple questions: Do we have any sharp objects, drugs, or alcohol, or any recent injuries on our bodies.

If you did, then this would mean getting strip-searched as this dog was about to smell our whole group one by one. We opened our baggage so he could have a sniff at it, and just after that were told to pull our bags ahead a few feet to hop back on the bus. Our search being part of the two-hour drive was a first surprise military order. Should you have any of the just mentioned, they would haul you separately for a strip search shortly thereafter.

Well there was only one fellow out of our whole group that was taken away. We finally arrived and were welcomed to Boot Camp and told we would be sent one at a time to our sleeping quarters in front of us. "You are # 8 Platoon."

The building was divided into four separate parts and the officers called each group of thirty men and women's squad number. "Kindly remember all of this being part of your Private role in graduation. Adjust your clocks and wrist watches." We would be given a bed number and told that wake-up time would be five a.m. – no sleeping in. Bunk beds in each two-person stall, with a locker along its side, and a set of blankets with sheets on top. Soon, we stood by the bed for our first head count in the sleeping quarters that were to be our new home. When they called the numbers you would say, "Here," to later be informed of all the training orders, to ensure all aspects were covered. From food consumption to lockers, uniforms, shoes, and gear as every item was to be marked with your family name, including the platoon and squad to identify everyone at all times. The power of a magic marker was part of ratification or else issues could result in fines – all coming off your pay.

Nova Scotia was the home of the well-known Mi'kmaq people, who had invited the European settlers to use the soil, fish these waters, and of course enjoy life in this new wondrous easterly coastal area. If you want to, you can step across the threshold of history at Port Royal National Historic Site, featuring a reconstruction of an early seventeenth century French settlement. Costumed interpreters demonstrate things like recreation in one of the first settlements in the New World.

In WWII, this Boot Camp area was the English part of disciplinary measures for soldiers training in the military trade required as part of self defense. Of course the location was ideal as an east entrance way. Recognized as having good and practical uses all over the world, my suggested trade choice was, "Chef/cook." I was told it always had openings no matter where you went. The food varies but is in big demand anywhere you go, so get ready for your departure time in the middle of October 1975, just two weeks after my seventeenth birthday. When all was said and done, I followed the rules regulations of the process, thank you.

Here I am once again, spending time with parents in thanks for signing the permission sheet of their legal entitlement till I was eighteen. This how the rules and regulations of government for its military system worked. Living in a farming area off the Trans-Canada, my mom smiled and stated a hope that I would still achieve my grade twelve. Big hugs to parents and off I go for my first plane ride all the way back to Nova Scotia only this time from the Halifax military base area to the infamous boot camp of the well-known CFB Cornwallis.

I'm sharing my story in writing this book to propel others to join, or to at least listen to others' time spent in the CAF for a short bit of time, or till they retire.

Located on the small peninsula flowing right into the Annapolis Basin, several miles east of the Bear River's mouth playing its role of A to B of the railway division, a gypsum ship-loading facility was located on the southern shore right in Deep Brook, NS. The east coast mainland's relatively level by the quickest tide in the world site that I had proudly joined to be part of the English speaking division. Here they trained recruits who were destined for service with one of the three operational environments of the entire Canadian Forces (land, sea, or air). Two floors of sleeping quarters were divided into four separate squads of thirty men each with good size washrooms, and an eight-foot floor space. We slept on two metal and mattress bunk beds divided with each other's lockers and a personal locking storage bin. Two large cafeterias for feeding times accommodated approximately 1000 people. As a new recruit with his parents' permission, I experienced the military disciplinary measure of doubling up with a full kit on the graduated ten-mile hike of Cornwallis's Heartbreak Hill finish line. This was to prove that your skills and training had been adequately to be used in our daily lives as civilians and most importantly, as a good soldier. There were indoor halls for women and men so the platoons could march for graduation practice on a cold day.

With big time historical value as a CFB base till 1995, the base is now in civilian use as a public retiree space, still open to cadets. It's now open to the public and has a park with a small industrial section, for a profitable future. As a former CAF base with the historical United Peace Keeping Mission, it was sure well liked. Nowadays it is no longer what it was back then, due to government minimizing military expenses right across Canada. It's memorable to all the fellows who were trained there back then, in military team style, not in competition against each other. The many adventurous, brave ladies also had two nearby platoons. Number one rule – we are all treated the same way from the start. The 3 S's at the start of our day were difficult to do for some back then. I was with #8 Platoon which was broken up into squads. I was in the third squad. With yearly openings for platoons upon your arrival, the ladies did the same training as the men. They were told to stay sharp in a respectful way at all times – no messing around till we were posted elsewhere. Stood in a long line to get the brush-cut hair that day for men. It was a wee bit longer for gals. In a humorous way, the long-haired fellows were only halfway done then told to sit down and wait on the side as we stood in line at the barber shop that day.

Two years at Algonquin High School, learning to play the trumpet had paid off; by God yes, I got to wear a Scottish highland kilt with my underwear on for a breezy day. Practicing as part of the music team in our camp was so cool. Best of all on my week-one arrival, I was told respectfully by the dental team that I was a chosen example for complete dental care for my time there. A half-dozen trainees stood around me that day looking into my mouth as a volunteer described of my severe extremity. Many of my teeth were like rotten wood they said, but worth repairing as here in the military dental office they referred to me as a "code red" in teaching, to the first year apprentices. I did get to spend a few extra weeks on the base to comply with this complete dental care, for which I am fully thankful to this day with all of my teeth still in place. Left, right, left, yes I did learn to march by command in a very impressive way, of course handling weapons in self defense along with learning survival skill to indeed last a lifetime. Did so till just a few days before X-mas out there, for a total of nine weeks in, with a few more to come to fully complete, and then be transferred to my cooking trade choice. Full of proud moments sharing the experience in being part of # 8 Platoon # 3 Squad was. From day one, listening to the words of command to be a good soldier was memorably worth the while. Dress code was very strict and training for survival skills was physically demanding.

Safe rifle handling for one another out at the range even on cold winter's days of twenty below. Our ten days spent outdoors sleeping, sure made good use of our own five-star sleeping bag. Pack-sack kits for both women and men had the basic survival food supply. Stories that had been told to me beforehand by relatives and other military personnel of North Bay all turned out to be true when I was living the life of a soldier.

My parents sure were proud and impressed by me doing this and gave me a salute upon my arrival to our X-mas supper. Family members at home looked at me with big smiles that were A-One.

Well, one week later, off I go, heading back to Boot Camp to complete my certified military training adventure. This was the best of all. We were then in the last few weeks of the January winter months, for a ten-day survival test at a nearby area, playing war games followed with a ten-mile walk/run from that area. Carrying a total of about eighty pounds in gear with a pack-sack and rifle in hand all the way back to Boot Camp over a floating water bridge of open water with broken ice all around at the finish line. It was watched at both ends just in case you fell.

Heartbreak Hill was the last part of the ten-day survival to properly be used in shaping us like a piece of metal on an anvil. The military training in Boot Camp had been modified to an efficient eleven to twelve weeks in completion after that Christmas. Our certificate came with a memorable presentation of a march, of course with a salute at graduation. After its completion, pictures were taken for you to keep, along with a crest. That evening we were allowed a cold beer to enjoy this celebration. A few choices were posted elsewhere for trades. Our proud graduating platoon

ID number was # 8 platoon. I was in # 3 squad with an individual soldier ID number of importance to be kept on personal items and in your memory at all times…or else.

The following day was a bit of a surprise, however. We were told by our commander that we were ready. As the #8 platoon were the most recent graduates of this CFB Cornwallis, we were told that our training and skills would be put to work on "Guard Duty" The winds increased for a period of a week just before "Groundhog Day" of 1976 and there were winds of 188 kilometres per hour. House roofs were pulled off and nearby trailer park units had flipped over. Loose material scattered about as we took turns carefully walking around the nearby communities till the wind settled. That last week before departure, was my last innermost dental inspection of my shiny teeth – good for until I grow old. Well, then they flew me out to the well-known CFB in Borden ON for the next step of the march to a six-month training of my trade choice of being an army cook. To do so anywhere in the world worked for me. We were congratulated for of a job well done, as we prepared ourselves in registering for the course, to then be followed with a posting in another part of Canada, then afterward to other parts of the world. Sure was looking forward to this part. Rules and regulations had minimized to lesser level than in boot camp, however, taking orders was still part of respect for the day.

There was the motivation of being promoted to a higher rank with better pay; from a private to corporal, in your trade choice. Now I was allowed to lead life as an individual and allowed to dress in civilian clothing, still with total respect to never forget you were in the military. As reminders there were fines for improper haircuts and breaking the dress code, and extra duties if you got in any sort of scraps. This was part of the troubleshooting reminder to discipline. Then the next step of course was that of dishonourably discharged. If events did come to the legal process of jail time, it was stricter in the military than the regular public service. You could be charged twice, if your trouble was in public with any citizens. This was a reminder that it was too easy to enjoy life, kindly respect your military vocation and be dedicated in doing so.

Well here we go. As I learned my skills in food preparation, it sure was an enjoyable and well appreciated opportunity to do the best I could. Training with older men as well got me teased for being a French speaking, as my first language made it slightly challenging at times. However, the French cooking terms in the final exam made it all the easier for me after 5 months, to become top of my class. Secondly, for me the offer to be sent to Montreal during the Olympics as a cook representing CFB Borden since I was top of my class was another place where bilingualism sure paid off. This made me happier than even the bonus that was given to be the squad leader of our cooking group for a bit of time. Now this was a definitely a moment to be defined as "je ne sais quoi," a pleasant quality that is hard to describe.

All at once things were made difficult as I got teased for being a Frenchman, and was disliked by a few as the youngest in charge of the position for our class, which was military style, to be obeyed by word of command. This was a moment of tribulation that I would never forget. I had a few cold beers at the bar, listening to music, and sharing the dance floor in bravery, to nicely ask with a smile, a few lovely ladies to dance. A Squad #3 Newfoundlander friend of mine from Boot Camp, and now a co-worker, who had been born in the community of Come by Chance and was celebrating a soldier's birthday had just gotten into an argument at the doorway. He was looking to get into a scrap with the sergeant guardsman at the door. I tried to walk over to calm him down and he pushed me away and he kept showing off his closed fist over a silly moment of being kicked out, rather than being hauled to the drunk tank for a disciplinary measure that was known by all.

Well I should have been easier going. I sure tried as best I could for that time. Now it was the day after with my first hangover; one day after the gathering at the bar in celebration of all that was coming my way. Sadly, that evening, at our four-man sleeping quarters, I got into a bit of a scrap with one my roommates who insulted me then stated that if we were civilians we could settle this in a short bit of time. Well, as a wrestler at Algonquin High School weighing in at 200 lbs. and well able to do many push ups and chin ups, I offered to settle it right that very moment without realizing that doing so was going to interfere with the final result for me in the military. Well, we did get into a scrap, and this twenty-three-year- old, six-foot, 170 lb mouth piece gave me what for in the hallway. I punched him in the nose then had him in a headlock with a few of my classmates letting us be to sort this moment out. All of a sudden two security officers came out to stop us and make us shake hands over it all. For me there was an invitation the next day to our Base Commander for a trial on our dispute, military style.

When all was said and done, they gave me no jail time but I got a $100 fine and loss of my squad leader position, with remaining days on extra duties every of the remaining days and evenings till graduation. The worst part was that this also included the loss of my trip with a chance to represent us in part of the Olympics. That sure was a mistake and a lesson in how quick things can happen. There were a few laughs from my classmates as a reminder of my over-reaction to being a team player.

One of the extra duties was with my Newfoundlander buddy. After class we were dumping all the recyclable food to go to a local pig farm in a slop bucket-hauling truck. Going over our mistake of what had happened as we were carrying an extra heavy bucket, he and I fell into a 9 X 20 foot bin loaded on the back of hauling vehicle that came once a week. With the biggest laughs ever as we walked back to change. Upon my graduation day, I nicely asked to be released from the military section thinking that perhaps I might join at a later time in life. Yes, my lesson was learned and I was thankful for the opportunity. Being still legally under my parents' name I headed back to the farm in the country.

Thirty years later I ran into the gentleman I had gotten into a scrap with over silliness, and we shook hands.

CHAPTER 8

LOS ANGELES, CALIFORNIA

Hitched a Ride to California 1977/78

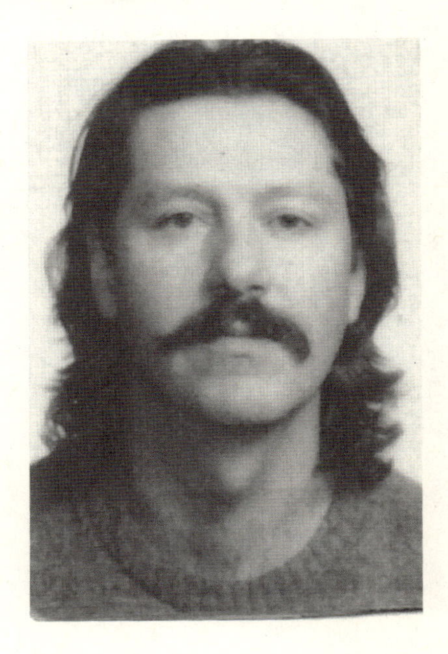

My first passport picture

As time went by, work as a labourer combined with skills learned at Algonquin High School were later all put to use with building tools. As a good worker, I got offers from our bosses, who were often Italian contractors, to cut the wood, pound the nails, and use a measuring device, as well mix concrete to trowel and finish it off for him. Back in North Bay, here I was back on the wheelbarrow as a construction labourer by my dad's side. Cleaning up as if going through or part of the last steps of a few school classes with add on, houses building, my initials were left written on the concrete sides of our City Hall during this busy year. This particular summer day, as we cleaned up at a brand new hotel building we sometimes worked indoors on the rainy or slow days. Like at many other sites, there's a final clean-up just before the contract is done, till the next one. Rolling around with a load of concrete floor slabs that gave the floor man time to finish his part, I was familiar with it as we talked on the break. I told him about being on the fiftieth floor of a building in downtown Toronto back in 1973, carrying tiles as a helper, then later being a carpet installer. One day, while right across the road, I got to watch the CN Tower being built off the forty-fifth floor. That year I was one of the first of thousands invited to visit for free just before the grand opening.

Back to work, anyhow this day we were working indoors on the cleanup at the Ramada Inn. Once again, my dad ran into the Head Chef and told him about my military background as a cook. Well, this day I gave a one-week notice to Winston construction, for a Sous Chef position in this new hotel, along with ten others for a three-month term, to see who the best cook was. I shook hands with the Swiss Chef and of course thanked my dad for his good use of break time.

There for a year, Holy Smokes, I then transferred to one of the other branches in the big city of Toronto in the downtown area. I lived just a few blocks away from Carlton Street, by the famous Cash Box, Maple Leaf Gardens. That's where one day I got to meet a well-known lady by the name Xaviera Hollander who lived nearby, along with a few other well-known folks.

Well here we go, heading on the second year of food prep at a busy hotel on the twenty-third floor, just around the corner I got to see and cheer for quite a few of those great hockey games. It was still required for me clean the fridges and haul the garbage down on the elevator as part of attaining my required hours for the "Red Seal." The same Swiss Chef had taught me how to care for all the important aspects of food prep going on two years now. It sure paid off big-time with a European touch on the exam. Working with many ethnic groups at this very busy popular chain of hotels was great, laughs at breakfast working with a Jamaican crew of three. Yeah' man, to blood man. Got to learn some of their dialect along with their style of living as well a few of their dance styles at a local pub in TO.

Here now the working with the Jamaican folks, then met a chap from British Guyana. As we worked together he told me of a few family members he had, who lived in the U.S. and he was moving back down to LA himself. His younger brother had come up for the visit as he had a place and was waiting to show him the way to drive down there with the opportunity to move and live in a warmer climate. Here I got the offer to hitch a ride with them as a Canadian and was told if you find work under the table without hassles for seven years you can get the dual Canadian/American citizenship. Huh well, this adventure sounded good to me at the time, by golly. Two weeks' notice on my second year with a few dollars in the pocket, notice to my rental, and one packed bag ready to go. Yes, I told my TO uncle where I was heading to, just in case of emergencies as a precautionary measure and I nicely asked him not to tell my parents as they would be upset. Told him I would call as time went by, and off we went to LA.

Sure was a memorable trip with my welcome to the U.S. being a full body strip search at the border crossing of Windsor, ON. That is part of crossing the border for all as a government safety protocol we were told, and is now still done.

We had departed just before Christmas and it was an awesome ride offer through seven or eight US states. Our arrival included a whole free month at his apartment in the black district of LA, before his wife and children moved down from Toronto. Well, I tell you, it was different from Toronto, Canada. I was not the only white man living in that district and was nicely treated except for one day when walking by a local bar, I got a knife pulled on me then asked for my wallet. When it was opened they found Canadian money and then asked me if I was a Canuck. When I nodded yes they apologized and handed my wallet back to me wishing me a nice day when I told them I lived just around the corner. Welcome to the neighbourhood with a cool hand shake and then an offer of cold beer.

Yes, I did get to touch the Hollywood sign – sure was part of the adventure for me. After my month I had to find a place of my own as that was part of our travel agreement. Shook hands in thanks and was told to call them if needed. Enjoy life and good luck. Off I went to the downtown area of LA. While later living on the streets and standing in line for sleep quarters as they do for the soup kitchens, I volunteered at these places, donated blood at $9.00 U.S. a pint, and asked the Canadian Embassy for bus fare for my return but still was on a waiting list. Well, one day I did meet a Mexican fellow and mentioned being a Canadian. He told me of a place that did some under the table hiring for experienced folks willing to work quietly for cash money. Invited to a Mexican supper, I met another fellow who took me and introduces me at a very busy restaurant, where showing my Canadian ID got me hired right away. With an unbelievable spark of inspiration about becoming a US citizen in seven years or so if well-liked by the owner, I prepared food at a busy twenty-four-hour joint for cash. It was all good with no worries I thought. I had learned a bit of Spanish at Algonquin and did get to use some with the Mexican fellows. Wow, all of sudden one hurried by with a few moving around quickly to leave. When asked what was going on. Ay Chihuahua, I was then told that as an illegal immigrant, the rules are to hide or if you get caught you then get kicked out of the U.S. Because the shadow economy could get worker and owner in trouble, the job requirement was to quietly disappear when an investigation of your work place occurred. This was new to me. The path of citizenship had unfamiliar rules and regulations to be obeyed or else. Via con Dios muchachos, good bye.

That day I was wearing a helmet and riding doubles with my Mexican buddy on his moped down under the highway storm drain system on a nice sunny day, going to the downtown LA of Little Mexico by invitation. For a couple of days at my chum's family place, I was told that in the white-man's district they had soup kitchens, along with line-ups for sleep over availability. There were just a handful per evening and the secret was to eat first then to go stand in line. It worked very well, oh and keep your shoes on or else. Got to sleep on a few less travelled park benches with my clothing bag covered with a newspaper for evening sleep. I asked myself what in tarnation had brought me here in the first place. Oh yeah right. Adventures of mine. I looked in the sky and said, good God.

Not looking to do this for the rest of my life that was for sure. When all of a sudden, standing out at the entrance of a sleep-over facility at the start of my day with a toast and coffee breakfast, I swear to God in thanks that an African American pastor of the area, walked right up to me to say hello. We shook hands on his approach with an offer to sleep over to help him renovate his church. He said he would pay my bus fare back to TO. I talked to him more to make sure he was not lying, and I checked with the soup kitchen and learned he was well known in the area.

In my three weeks at his church I had a nice bed to sleep on of course fed very well . As I got to meet his wife and three daughters. "You see son," he said with a smile. This was one of the big reasons he had asked me to help paint and do the flooring, along with heavy lifting of the benches so he could have his first mass at the end of the month. Later, he bought me a ticket and packed me a big lunch just the day before Muhammad Ali was in a match. We listened on the radio that evening on my way home. Sure enough my Godfather Viny was there to pick me up for a break, then sent me on the next bus back home. My parents were now living in North Bay and they picked me up at the in-town bus stop. Then as I was broke, Dad drove me to a past tenant of ours, a great house builder from our Bonfield hometown.

"We're off to Timmins, Ontario, a mining city. It must be like that place where your brother is at, eh? Your dad told me you're good at pounding nails and cutting wood so you'll get cash pay for three months hard work seven days a week, till we finish the whole street in that city, alright?"

Sure enough, this was straight-forward prefab housing. The walls were pretty much there and we used the measuring tape to follow the lines on the floor you made to nail them together using the level. Showed me how to do a couple in the proper way then told me it would be easier next week as we had a half-dozen other experienced workers to pitch in.

Right on sounds good. I asked him if they were local fellows that he got through Employment insurance for openings, since I knew about that from my work experience.

"Well. Sort of," he told me. "Hop in the van and tomorrow I'll show you." We drove a little ways out of Timmins, and with a wonderful view of the area, we crossed the bridge to arrive at the community of McMurrich at the Monteith Correctional, to pick six men out on early release. Oh well, I was told these fellows had run into a little bad luck with scraps, or drinking, or driving badly, or something silly and that they had now paid their debt and were looking to get back to work themselves. So our crew would be three fellows for each of us with years of experience in construction. So this should be a breeze – we just provide transportation both ways for the summer. We would hear some jail stories from these fellows and as a few spoke French, our mother tongue, we should have a few laughs as well. So that day after the paperwork was in place, I got to meet these fellows as they asked us a few questions about the work area, along with details of pay. Overall this was a standard day's work of ten hours before the return to our temporary homestead. Well, each day there were inspectors who came by along with watch guard supervisors of the Monteith site to keep a head count or so I was told.

"Are they all still here?" was said with a smile. At any rate these houses were built on a time frame in more ways than one. Straight up and levelled evenly, which was verified by big Mike, our boss, for a payback of cash with one cold beer at the day's end for all the hard workers. Minimum wage was beat by a few bucks more than the five dollars a day of many workers in Canada back then. This worked well for everyone and I later got a ride back to North Bay to visit my parents before my departure to Yellowknife, and had learned to save a few bucks.

CHAPTER 9

HALF OF MY LIFE IN THE 'KNIFE

Yellowknife, 1979

My first visit to the real North. Yellowknife, NT 1979

Persuaded to quit travelling to go live and work in one place at least for five years, and get your grade twelve, eh? Listening to my mom and dad sure paid off, to that I testify. Working in a few places was proof with yet another movie clip that my life was truthfully an adventure. I never knew about making the news, a radio show to now finish this book!

Your brother and his soon-to-be wife are way up north. They have a room for you to live at until you get things in place to get started on your own. Your brother in Yellowknife is working full-time as an auto body man, and he told us there is lots of work in this mining community for those who make it there. Worth giving him a call as he told the story of how he got there in blink of an eye. While living in North Bay without a job, being a pool shark he played pool at a local hall against this man and had beaten him well. This fellow was passing through on his way from Toronto, because one of his favoured pit stops was in North Bay ON. Being from the "the Big Nickel" of the Sudbury, Ontario area, upon his loss of a few games, for some cash he offered my brother a ride all the way to Yellowknife, where his dad still worked in the mining industry. To a dreamer's delight, there were supposed to be quite a few mines up there looking for workers with availability in all trades. I was reminded to dress warmly as it is very cold.

That following day my brother hugged his honey and called our parents as he packed a bag and off he went with Ron. Hell of ride of over 3,000 miles. One of the offers was to be a designated driver to make it there quicker. With a stop-over half way through, they made it there in record time on the fourth day for a needed rest. Ron had told him that he worked at the jail in Yellowknife and showed his ID card. Yes, I believed him and sure enough my brother got a job offer right upon my arrival from one of Ron's good Yellowknife buddies. Our Lord works in mysterious ways, so pack your bag as he sure will give you a chance to use the many skills you have learned so far. We have heard that the cold temperatures seem to be a roadblock for many up there, however, the summer has the twenty-four hours of sunlight. Sure can't be beat.

Sounds good Bro, see you up there.

Hugs and handshakes to get on the bus to shoot for full time work was all okay. For me. Parents once again stated for God's sake try and complete your grade twelve while you're up there. Eh! So off I went with a big hug and a huge lunch box for the five-day bus ride. My dear, loving mom had clean clothes nicely packed in order, with a towel and seven pairs of underwear with as many socks –well for a week. Never better part of planning. Sure was a long ride, chatting with other travellers and putting to use reading material left behind by someone? It was free, and some of it had good update info about the path I was on. I had no idea while heading to the upper region of Canada for work. About the growing profitability and potential future resource of its great wealth in mining oil and gas. The Canadian Forces Northern Area -HQ where I later got to work at as a Commissionaire certainly filled my pocket in pride. I had never thought to see the CFS of Alert base in Nunavut's northern region. We had humorously heard that that such a posting would one day be offered to us as a private.

A part time position for none other than Buffalo Joe's air service while covering a golfing buddy of ours for his courier service drive of three weeks holiday time, out in the Great White North. Defined by the Standing Community of National Defense, our northern region of Canada is neighbour to Alaska. They are both part of the vital use of the northern passage sea route. Transportation to it by air is out of Yellowknife, NT's capital. CFS Alert being so far up north it was humour to think about getting a posting there and well, now I knew why. As NATO connected us to North Bay's underground base for a good fifty years that is still part of the northern sovereignty shared with our American neighbours and is now also part of the Manitoba section. They use to give tours at this underground facility in North Bay, to show students, tourists, and public visitors the CAF's massive ability to confidently look out for us all. With its technology in place for Earth's need to help one another, I was on two tours; one of them part of a school trip supervised by security as they toured you around. They no longer allow that. The second was with family friends in the military who had never seen it themselves, and said let's go check it out.

So off I went to the Knife 1979.

Adrenaline flowing, with a chance to redeem myself against the statistics, off I go way up north into the Yukon. Travelling across five provinces, it was my longest trip ever in mileage anyhow. For sure it was my first time way out west of Canada and I got a quick sample with a great view at times, in between stops to pick up others. It was a long bus ride to say the least – affordable overall at about $100. Two thousand miles behind me, here now arriving to Alberta, yet a MERE thousand miles away for a twenty-four hour drive to the Northern Lights. I was looking forward to the chance to have new experiences and live with northern cultures of the well-known mining industry of today. There were only a few of us left on the bus for the last 500 miles from High Level Alberta to the Great White North of Yellowknife, chatting with one another about heading past the Mackenzie River crossing on a twenty-minute ride on a Merv Hardy fairy vessel. Of the few passengers left I got to spend time speaking to a Dene family which had been up there for many generations. A husband with his wife and kids were also on the long ride out of Edmonton, Alberta to the Knife. He told me the bus ride's arrival should be at around 12:00 p.m.

This father was also part of the mining industry as he flew out to isolated areas farther up north for a darn good pay. Wee ha! Finally here as we hopped off the bus, tired as could be, with a smile and saying see you later and goodbye to one another, heading our separate ways waiting for our bags and what not. It was still nice and warm upon our arrival. Sure made me feel like I did real well.

Here we are at a quiet section of this fall arrival at the sixtieth parallel. I was sure exited to call my bro upon my arrival. After the long ride, a good sleep with a warm shower would work for me. Yes there was a nearby payphone and I had some quarters to make the call. He answered to say, "You finally made, eh Bro? I'll come to pick you up, so hang on for bit. We are just around the corner. Be there in five minutes."

Yes there he is with his new Camry. *Wow,* I was thinking as he pulled over by the bus to pop the trunk open as I dragged my gear to it. Followed with a firm handshake, we tossed it in. "Hop in for a quick tour," he tells me. "Then to the house that your new sister-in-law and I rent." With a quick round-town tour of the many new things, he points out his workplace and nearby hospital, hotels, bars, the theatre, and grocery stores, all in a short bit of time. Told the population was less than 5,000 people in this busy mining community of Yellowknife with folks from many parts of the world who speak accented English. So off we go to their home for my desperate need of a shower and a good night's sleep, thankfully finally arrived in the Great White North. "This being a weekend you should get a job by next week. They are always looking." As we walked in, he showed me into the spare room then the shower. "See you in the morning after a good sleep." What a great sleep. Best ever with dreams of amazing prospects coming into my life in YK.

I got to meet Aboriginals from many parts of Canada, along with local Dene folks who had been there for quite a few generations. Inuit folks were also around populating the area as some came from the northern areas of Nunavut to achieve their grade twelve and give birth to their kids. Most impressive, in giving my family name I was told that it had been here for a while, spelled the same way as ours who had been there for over three generations themselves. After my genealogy research of the same family name, I was thankfully registered as part of the North Slave Métis people in this region. This sure inspired me to finally achieve my grade twelve education, learn the PC, and be part of social cultural development as a participant to many Aboriginal events. Ambitions were well worth having as they were not cheaply won was true for me. Being part of the many local events and competitions was always worth the while.

One of them was to help build a steel teepee with mining drill rods for Association Franco Culture in Yellowknife near my shack just off Ragged Ass Road. After my first year of living there, I was taught winter survival of net fishing with hunting and trapping as they had been done for generations like my family did back in Ontario. As the Métis had come since the 1700s to be part of the fur trade in this region, our family names from both sides was part of the surrounding communities. Mr. Goblet whose wife and son I worked with at the Yellowknife Inn, would offer invitations to his home for a wild game feed with the skinning of his trap line off the Beaulieu River of the Great Slave Lake not far from a mine site that I had worked at with one of his other son. From the many families who came to this region to meet the high demand of the mining industry since 1897, I met a few Ontario, French-speaking folks from our area who had done so as well. With accessibility by air, rough gravel road, and railroad to Hay River, NT the population had grown to accommodate the demands of gold, like the Yukon days. Like the many miners before, who had come there for reasons of good pay, along with training for the many underground positions that were always available till the end? No offence to the industry, I was a surface man after the visits to a few mines old and new. There was no way of being underground for me, even just for two-hour tour. Shortly after my arrival I started looking for work and there were many incredible mine-site positions offered. These were big time on the board at the EI office in Yellowknife. Being a good listener on the streets and while in the pubs for a few pints, I heard about promotion possibilities with free training. For the many single miners to meet ladies they were sure always looking for more workers, along with music and dancing with dating in this neck of the woods. When hitchhiking on the highway in the -40 cold, you legally have to be picked up if not by a passing-through stranger then the RCMP may get a call to do so, as a precautionary measure for northerners. As it was a little on the rough side now and again for those who had a little too much to drink or would just get kicked out with no discrimination at all. Well respected precautionary measures of the RCMP were there for public safety to all – the drunk tank was used twenty-four hours a day, especially in the winter time.

At 40 below, you could slip and fall and could die within the hour. It was like the gold rush days of the past for crying out loud. Here we are almost in the 1980s, with it being like the Yukon Gold Rush days and then some. Staying at my bro's now, and my new sister-in-law's they took me around town and updated me with the do's and don'ts. Got free underground tours as part of an invigorating motivation to coax newcomers into the mining industry.

Great pay persuasion for the long days though physically demanding. There were jobs for men but also availability for brave women. After my exciting tour from EI for prospects, it was just not for me. I had heard from a friend or two theory only added truth of what you would hear at the bar or on the streets by friend's relatives of a growing town there was a lot of work all over Yellowknife. My first week of 1979 was so memorable with proof that work was here.

The usual cold temperature in Yellowknife, NT. 2009

Not long after, on a Sunday morning as I sat at the YK Inn chatting while having toast and coffee, a fellow with his wife and kids who'd been in YK for five years gave me an update overall. Workwise, he told me he would speak to his boss today about me looking for work. He said he was sure of my being hired as a builder's helper with him, building houses, garages, and whatever. As a hard-working Newfoundlander builder he was aware of the expansion of the now-growing mining community. He told me that I might only get a wage of ten dollars an hour to start. Wow. Sounds good to me, sure would like a chance at this job offer.

Next to Deh Cho bridge NT. 1988

Within the hour he called the boss and over the phone, I was told I was hired. The next day on my first week starting, I got to meet the boss with a hand shake and filled out the paperwork. They told they would give me a trial till fall ended, since construction slowed down at 40 below. I was to help out my new co-worker on ten-hour days. He was from the Bay Bo'y, a very well-liked, skilled Newfy carpenter by trade who knew about many things. Asked have you got a valid driver's license, which I then showed to him, "Yes Bo'y," he said. "You're good to go to pick supplies." We slowly started with a tour around town, so I could know my way as he showed me the how, where, and when of material arrival. Then upon its arrival from the south of Edmonton, all the way to Yellowknife NT, to lead the two-piece load to the lot address, pointing the right way to combine the second half of the total 22 x 66 foot home being installed. The trucker then pulled along the side of the two-piece mobile home we would be then put together within the crane lifting pieces of heavy equipment. This was sure something else – the first time for me and I enjoyed every moment. Amazingly, I was told the folks could move into the house a couple of weeks later.

There I got to meet a few of the well-known ice road truckers, including Alex Dobogorski. A few weeks before he had been working on a huge caterpillar piece of equipment near our prefab lot location. Yes, that day I worked alone outdoors on a cold winter's day, nailing away on a garage unit for the prefab homes. When all of sudden just out of nowhere a big fellow, also dressed in his winter gear, ran through the buses towards me in somewhat of a big hurry. Out of breath and standing about five feet away from me, he smiled and then told me his name, then asked if we had any toilet paper out here. "Yes sure. Help yourself." It was there right inside the truck, with a plastic, five-gallon pail poly bag with a Styrofoam seat at the back. "Nice to have met you," I got to go and off he went back to the bush. Anyhow, as it got colder in the Great White North, an indoor position for my first winter sure made sense to me.

Start of my year 1980 I was working at the Yellowknife Inn, an all-around cook feeding the incredibly large groups of the YK INN. Like no other busy downtown, this particular hotel with a dining lounge that seated 65, served twice as much on a busy night. The main attraction was a view at the Dining Lounge entrance-way, with one of the largest polar bear furs in the world, which later got stolen, oh well. Right next to it was the main entrance-way with capacity for over 120 at a walk-in, cafeteria-style buffet with a morning line-up every day at opening. Hotel capacity was over 100 rooms, and at this particular time, it was the busiest place in town, for sure. At the side entrance-way was a large banquet room of over 250 capacity. It was considered number one, for many special events; weddings, sporting events, X-mas parties, St Patty's day…the list went on. Aboriginal events of all sorts, from government meetings, to dances for celebration of life in the Great White North.

A Dene event in 2001 in Detah, NT

Sure was a great suggestion by my Newfie co-worker as winter work availability slowed down for a reason, indoor work had options, with all-in-one accommodations as part of the workplace deal sounded good to me? Long shifts using my military food preparation skill with the offer to be their future Sous Chef in the kitchen at this busy place covered all bases. The day I got hired by a Canadian Chef extraordinaire, we shook hands and within five minutes he told me "It will cost you a box of beer after shift, you start now." A few years later a British gentleman did run the show he sure liked my army-style in food preparation to have me create a full menu.

They had a guard at the entrance-way door for the huge line-ups, sure something to think about as it was very demanding in all aspects, but I indeed enjoyed all my time spent there. A big surprise one early morning when to fellows entered the door with guns followed by a fellow behind them, showing me a badge telling me they are beating the crowd they asked me for bacon and eggs for them and Mr. Pierre Trudeau. I did continue to work the long hours catering the demands of food service at the Yellowknife Inn and living at the awesome, free accommodations as part of the deal. It was a party house for all staff, and sure was worth every penny as it did allow me to go there on breaks, which was sure a big need. Living in a small community downtown had a lot to offer as it was walking distance to the rock and roll bar of the Trap Line or the Mackenzie Lounge. Across the road the rowdiest pub ever named, Gold Range, aka the Strange Range country bar was just across the road. With secret, big-time gambling held on the second floor that I was allowed to watch for potential interest in being one of the players someday. There was a strip joint endeavour, which was also well liked, slightly hidden to the side in respect to the ladies to peek or not.

Putting their hard won dollars to work for the early retirement attempt seemed to be the intention of many. Two jobs were held by a few to make this dream come true along with a little gambling at times. It did work for some as they won at poker and also invested in stocks and the market shares, which was also booming.

Some sold there houses because of a tip to make a fortune in the outcome.

As time went by we got meet each other at more favourable events as we both also drove taxi as well, and played soccer against local inmates for fun winter events. Alex bounced at the Gold Range sometimes, what a guy.

The town grew, I tell you, there was no messing around on a ten-hour shift with a choice of six or seven days a week, which my boss told me when he came over on my first week. I got a raise of a whole dollar, and was kindly told to slow down just a wee bit, to a steady pace, and take every second Sunday off. You're not working at one of the mines for a bonus – the hard-working miners do so for a reason...the big bucks and early retirement and it is a very demanding workload. With my first paycheck, I took my brother and now his wife to an evening out on the town. They had been there for a few years now, and they knew that Yellowknife was a busy mining community with workers from many parts of the world. Of course there were Métis, Deneh, and Inuit along with a few Canadian Aboriginals; folks from other mining areas of Canada that came to work then brought their families. Mostly they worked in the busy gold production industry with bonus pay for those who were better at it, and promotion options. From one site to the other there was competition for those skills as the winter was demanding and treacherous for some. This was proof that minerals still held a big interest since the seventeenth century of the European's arrival, and even more so now all over the world. The Klondike Gold Rush days of the past were still in play, ninety percent of it still out there, or so I was told. You would hear stories in bars and restaurants of a mining town with picks and pans on the walls as reminders, and I was later in the Canadian gold and diamond rush myself. Although now, in today's time stock market manipulation was and still is in place to cause strife with others over town issues. Parents were right about the Great White North it was finally time to grapple with the difficulties of life with education awareness. I heard the truth at the Rec Hall and Gold Range and sure heard it while driving taxi at closing time.

There were cold winters with the awesome view of the Northern Lights. A few of the people I met living in the old town, just near the Great Slave Lake often had suggestions. As it was difficult to find a place to rent, I was introduced to and talked to the right man. I later met the owner with a few cold beers for a chance to live around the corner of Ragged Ass Road, for an affordable rent in a small shack. Just make sure you had lots of wood tucked away with a five-star sleeping bag for the forty below winter months. Smiling at the thought of only eight hours of light combined with the bonus thought of the twenty-four hours of sunlight in the summer.

MÂS'KÉG MIKE's Adventures-1980 to '82 continue, having met a fellow from Orangeville, Ontario while driving a taxi on the night shift in competition with another thirty drivers to provide required, well paid services to Yellowknife, NT. Folks often asked if it was okay to ask for my car's number. Yes it was. Call me on a personal. With no bush runs yet really in place now in 1982 working at the Yellowknife INN hotel, driving taxi part time on the evening shift, I got to meet a long-time friend from Ontario, aka Ace/Skinny Cudney.

MÂS'KÉG MIKE got to travel and fish on the many rivers in various parts of Canada. I nearly caught the biggest pike ever with my good buddy Ace out on the Yellowknife River system along the trails. We met a miner whose wife was a school teacher – the man called "Mad Dog." Like the well-known Canadian wrestler I had met myself on our well liked Algonquin high school's wrestling team. He told me he worked many twelve-hour shifts at the mine and that was why they gave him that name. Skinny was a day trucker – Ace, the taxi man by night vs *MÂS'KÉG MIKE* who is also mentioned in his book called, *As a Matter of Fact* by Skinny. I have brought back some wild game and sometimes fish for the feed.

After thirty years we still talk about bringing his sons to catch some fish and then playing crib afterwards as we have consumed a few cold ones together after a long shift with a break in-between, to sleep at his two-room uptown house just kitty-corner from the Salvation Army, which run by some great folks from Orangeville, his home town. Because of that he named his fun dog Sally Anne. After one crib night she had six pups right under my rental bed.

Anyhow back in the Knife in 1982, here now driving taxi in early spring I had met a young lady from the Quebec area who did not speak English very well. She had just moved to town as her brother had a place for her to live, and he got her a job at a small pub. She had also been given a nickname herself – "T" as her French name was Therese. The English found it hard to pronounce the right way and she'd then have you retry for the proper pronunciation of it. Working at a bar a name tag was often used in a noisy work environment for quick identification, along with a few other aspects of folks with the same name. Your quality work skills might sometimes be the reason your name was given to you or for another reason. *MÂS'KÉG MIKE* was given to me because there were many Mikes living or passing through this busy mining work area. For some their reputation was the ideal way to re-baptize them, and they might be named for the high traffic work areas, or one of the ice roads of mining in various parts of Canada. At any rate I did not mind it. In my early twenties. I was oh sometimes called Frenchy as this was my first language. Sure felt better to be called this name as the military style food prep of isolation was part of keeping a high standard. Should there be many other French folks that live around you it is better to just have a bush name. Overall to say when I was in the English army with my accent, I was called Frenchy there was one per squad.

In the early days names were changed or a second name was added on to individuate. Sometimes misspellings of names occurred due to lack of education. Having opened a bank account at a few banks over the years, this one in Yellowknife was just like the one in my home town. It had a few other fellows with yes, the same first name, also the last name. Piece of advice from the bank I got was to add an "a" to my first name along with my middle initial as a precautionary measure for there were twice as many with my name in the U.S. So like many others a code name was employed so as not to suffer from a mistaken identity in a small town of 5,00.

CHAPTER 10

ONE FOOT IN THE BUSH,

The Other In The Taxi

One of the many taxicabs I sold. Yellowknife, NT. 1997

Letter to the customer who was in a hurry as she waved me down on a busy main road, then complained:

Monday 21, 2008

Dear Sir/Madame

The YK driver's license bureau and many offices have a sign that states in the case of failure to respect the human rights of employees, you will be asked to leave the premises. I have spoken to a human rights representative on this matter. I am saddened by the fact it was assumed that the incident was my mistake, without first hearing both stories.

As a professional taxi driver in the city of Yellowknife, I often deal with many difficult situations every day. When my personal safety is compromised or my human rights are violated, these rules exist to protect me even within the confines of my small work space.

In regards to the difficult passenger from 5617 Franklin Ave. YK, NT, who made a mountain out of grain of salt – for the safety of myself and others, I had refused to provide our service for this lady on a Friday afternoon 4:30afternoon rush hour with extremely slippery winter roads.

As I entered the driveway off of rush hour traffic around 4:30-ish p.m., the lady was standing too close to the side of the main road for me to provide a safe pick-up. I proceeded past her and turned myself around safely to exit the parking lot in safe and professional manner, as it is a difficult place to get out of.

As the driver, I pointed out the danger and the reason I did not stop exactly where passenger wanted me to. (I only asked if she was a driver so that she might understand better the reason.) The lady seemed upset and said I called her stupid, which I did not, and she told me she didn't want to hear any lip and to drive her to Nova Court. I even offered to call her another taxi. Still, I complied with the lady's request, and drove her safely to her destination.

As the lady prepared to exit the vehicle she offered $ 20.00 for payment. I refused payment and asked her to please leave my vehicle, and not to take my taxi again. She then exited.

Complaining to the office in regards to this matter was, in my opinion, a poor attempt to cover the short-sighted, insulting comments she made, and she turned things around to make them sound worse that what they actually were. Personally, after I dropped the lady off I forgot about it. She said she'd had a bad day, but this is no reason to ruin it for others.

If she apologized I could change my mind
– our customers make the paychecks.

We met later and shook hands. No worries. Life goes on.

Driving safely is number one for all, and public service is a priority in dealing with people of all age groups. Some are going to their workplace or the airport within fluctuating time frames. They may be getting the baby sitter, or going to school or many other events. Payment is sometimes subsidized if need be as part of the job, whether or not, is in helping the disabled of all age groups, watching the kids for Mom, handling luggage before and after arrivals, carrying groceries bags to the door, or touring around to events. Tips are well appreciated and it is courteous to do so. Cost and money earned is equal to the eighty-plus hours of a week's work.

Parents with children go to the day cares, then the work place in combo trips, or they may go to food stores along with bars, to cover all other aspects of this busy mining community. With many trips to the roadside villages of Dettah – Fort Rae Fort Providence all the way to Hay River, sometimes it was very long there and back for some. Right in town there are indeed churches to accommodate the religions of many from other parts of the world who themselves work in various aspects of the mining industry by contract or for a good hourly wage. Along with well-known folks in sports events, every year as a promotion, retired hockey players played against the RCMP team, actors, politicians, or musicians at the Frantic Follies Summer Festivals, or the Caribou Carnival in the land of the midnight sun of Yellowknife.

With out of town communities that also required year round transportation services, they did depend the now capital city of

Yellowknife to support in many ways. As the exploration and mining sites came closer to their community areas, with some great potential for its future, it is still part of today. Because of the crews from Canada and various parts of the world, many do live in YK and there are high prices for real estate along with the extreme lack of choices in local accommodations. Now the mining demand does bring many from their home towns, but in summer time accommodation is more available because you could bring your tent. The park has potential and I did it a few times right by Long Lake. Diamonds, gold, oil, and gas are in big demand and the northern Aboriginal communities are experiencing it.

Yes, there are bad days for folks of all age group especially on a rainy day, extremely cold weather, or at a theatre bar closing. The busiest times in town are amplified by a combination of hours, plane arrivals at the airport, and terminal time frames. Longer trips pay more than a short in-town drive. The customer simply has to call dispatch ahead of time. Planning ahead simply prevents difficulty. There are meticulous dispatchers for incoming and outgoing calls for the rush hours of the day. They help train others on the awareness of drivers' challenges that might disrupt the ongoing service. Usually it's drunkards without money, or arguments on topics like customer complaints for a service provided. They pay the same fare whether rich or poor – that is all about earning a living to feed families, and pay the bills, just like yourself. Thank God for bus fare if the community has that service. Famous or not should there be more needed, choices like limousines are usually possible by booking ahead.

Back on the trail for many hours, providing the public with taxi service and taking some time off. Dispatchers helped us care for our cars as safe public drivers. If a car was dirty it was to be washed, or no trips. Contracts were offered at the beginning to see how well I would do for starters. Once your reputation with a code name was Okayed, off to isolate mine sites I would go. Though goods are sometimes flown in, often the winter ice roads are also a factor of transportation for supplies; in itself a cheaper service than flights to the mines when it's done at the right time. Of course transportation to town included the airport. The ice road was blocked off twice a year for about twelve week's total.

Now with a new bridge in place, things have changed with more access year-round at a now very busy mining City of Yellowknife with over 20,000 in population. Back then in the early '80s winter time transportation in YK was in big demand, and the taxi service played a very important role for the RCMP and fire-fighting crews, followed by assistance to EMT, and for incoming flights with Buffalo Joe and other airways. Taxis contributed to various aspects of hospital care like nowhere else. Some days, five hundred dollars was made – sure was decent bread for great services to the nearby Dene community. In the training for all drivers' safety was #1, as this was a public service and there were no adequate bus services back then.

In the '80s I was working out the isolated areas at the diamond explorations, at the time of an awful explosion that put the community in desperation, to say the least. We read it in the Yellowknifer news.

Anyhow only one cab stand was open then, but later there were competitive opportunities because a second and third company were opened to meet the demand. However, the oldest city cab company had the most long-time, well known drivers and it was later sold to individuals as shareholders. With the help of my lady friend Denyse Simba of Niagara Falls and PC specialist Mr. Amerik from India, with of course our lawyer, I joined in on that. My fifty shares are now worth a lot more than the $6,000 I paid while at Tim Horton's, that day. The sale offer took place at Tim Horton's on a Saturday morning, midsummer of 1992 I believe. That Friday evening we had played quarters and crib at the Strange Range, and the owner, who was a good Yugoslavian buddy of mine, told me straight up that should I be interested in the purchase of shares, to just have the lump sum of money by Monday morning or else.

"How much?" I asked with a smile.

"Well, no lies," he stated. "A total of $300,000 dollars, keeping in mind my son will get twenty percent right off the bat. By this coming Monday or else. First come, first served, okay?"

Here we are that Saturday morning, with a rolled-up stack of money a foot high, for the share offers passed on by our dispatcher lady friend who bought one herself I believe, sanely allowing groups of three to make it easy. Of the fifty shares that were available only to those who wanted them of course, we knew there was a lawyer in place to confirm the deal. Some bought more than one share with twelve. I believe his son covered a chunk right off, as he was all in.

Due to the use of a pen cell phone followed by word of mouth, drivers popped in to verify that this was all taking place. Sure enough a share for $6,000 dollars on that day was all true. This was at Tim Horton's in 1992 with our legal representative said it should be called City Cab 1993. All the dispatchers kept their jobs in fairness, to the public service that day.

With a bottle of champagne in final agreement, and witnessed that Monday at our office with a few of the shareholders, we shook hands in celebration of this event. Of course there was City Hall registry of the business name due to the local bylaw requirement to do so. It was appropriate to cover all the business aspects. Safety was a number one priority at all times .

Before that big sale of the 1993 City Cab Company, as back in the early '80s my good friend Ace a.k.a. Skinny and me, like many others were full/part time cab drivers. Often as a second job, with good pay for the many hours. That is how we met! Skinny was at the Golden Range bar in Yellowknife at the two-car taxi stall waiting for a trip with the window down as we chatted with one another on a slow sunny night. Oh so boring eh! It was a day off for me from the YK INN.

"Do you play any cards?" he asked me. Army trained on the coast by the Newfie team, yes boy. Off we went to play cards at his place just around the corner across from the Salvation Army, where the folks that ran it came from his home town of Orangeville, Ontario. Sure why not? It could pay off with a little break and maybe enough for a box of beer. As it was slow, we shot the breeze, watched the sports on television, and stuck around for a feed with a few other cabbies who were also from Ontario. We called dispatch a little later, but it seemed it was very quiet so off we go to park the car then pick up a box of beer for a twelve-game match – winner keeps all at five dollars a game.

We watched sports and news, and played a lot of crib through the slow times, as well as backgammon for a few dollars now and again. We continued to do so with dispatch access, in case it got really busy all of sudden. No beers allowed; the RCMP in Yellowknife were very strict for a good reason...to save lives. Yes, we'd put the game on hold and get out there right away, sometimes with the game on hold for a day or two. After a busy night we played cards by the wood stove he had there. Sure brought memories from back home and we chased it with a cold beer after. Getting tired after some late night cribbage, I got the offer for a sleep-over at Ace's shack, in the spare room. Yes, his lab-mix hound named Sally Anne laid under the bed while I was sleeping, and she gave birth to six puppies by morning time. What a surprise it was for me on that Sunday early morning.

As all bars were still closed back then on Sundays, the special service of bootleggers of extremely high demand was well liked. Thank God they finally opened the beer stores on Sunday in some places. Driving taxi on those days, if you were trusted by a one to take a customer to with a thumbs up of the okay Folks of all nationalities sure liked alcohol and it being a small community it was known by half the town. Alcohol was not to sold to the under-nineteen age group, which is still the law. Ace and I drove hack in the evenings sometimes with a pack of cards and a crib board at a mid-town area, so if it was slow, like six cars waiting we had plenty of time.

Bar closing was like the gold rush days. There were good tippers and I met people from all over, some of them famous. Share a taxi with anyone when one arrives, because the drunk tank threat is for a real good reason.

Often the new to town were looking for a job or had found one and were waiting to be flown safely to an isolated area for better pay, waiting for the big lakes to freeze up in the extremely cold weather. Of course band musicians would spend a full six weeks at the Gold Range because the ice road availability was on hold till the freeze up. That was a wee bit of a slow, tough time for some. Yes flying out was possible. This would be the definition of Cabin Fever gone wild that would make the RCMP put extra hours in public service.

Now, thirty years later, a bridge is finally there. Sometimes famous musicians there for summer events or well known, retired hockey players played against the RCMP in competition in winter games as part of the Caribou Carnival. Got to meet a few while watching the games at the arena that was popular and that promoted sports for local's big time. Mostly in the summer visits I met a few of our politicians and prime ministers, along with our British Royalty family who came for visits and needed taxi rides now then. One evening at the Range I got waved at by well-known movie actress Margo Kidder, who was born Yellowknife. There were many well-known musicians at the summer festivals with plenty of other folks just there hoping to catch a big fish off the Great Slave Lake or hunt a moose or caribou. For hunters it was a dream come true. Yes, a bit of dancing at the pubs after a long day's work was enjoyed by many. There were country style bands and the pubs were filled with local folks and new faces and there was lots of dancing going on.

Taking a summer break for twenty-four hours of sunlight fishing had some moments now and again that involved hunting for caribou meat that would give you a long stinky fart. Folks would ask, "Are you okay?" A few cold beers may have been the combo culprits for the gastric mishap, but walking took care of it. At home playing crib, the weekly lease and gas payment were without question paid for by the twelve-hour cab shifts or the longer shift version if the owner was not there.

Us cabbies looked out for each other during busy times with a lot of out-of-towners celebrating whatever events there were or time of the year it was, with calls to the after-hour parties if you had cash or beer. And looking for that right someone to share the night with was in big demand year-round.

As my good ol' chum Skinny/Ace and I hacked away to save a few bucks, we would allow fishing time or attend local events with other town/bar chaps, one of them called Life. With a few of them I went on an incredible fishing trips, sharing gas and boat costs for a week-long opportunity that others just would not believe or realize unless they saw or did it for themselves. One time I sent a better than three-foot-long trout to a good friend's memorable wedding. Sadly I was not able to attend Mr. Groomsy extraordinaire and his lovely wife's great outdoor wedding at a fun relative's back yard in the downtown of Yellowknife, land of the Midnight Sun. After the flight back from Lac Du Gras off the Coppermine River where I pulled a five-month stretch, I would communicate with trailer-tenants, cabby-buddies to see who wanted to fish or caribou hunt.

This is the mining site in 1992 where I was on a live radio chat with Mr. David Miller of CBC in Yellowknife NT, where he then flew in to discuss a long days work at this isolated region area of this now mine site that exist today. MÂS'KÉG MIKE is in this picture with the crew of a dozen hard workers that later where in a documentary with the local Aboriginals of the Yellowknife area I had fed them a traditional fresh caribou meal, along with fresh or smoked trout caught at Lac du Sauvage that day. My father Percy called to let me know he had seen us on television while in Ontario, still waiting to hear from CBC about getting to see it myself one of these days

We took pics of the trout's and other fish. One I had caught off a secret spot of the east arm area, called "Hole in the Wall" with a great depth of water. When I thought I was hooked on the bottom, our buddy laughed and said you don't have near enough line on your reel for that. The water was deep and cold, even in the summer. If you were to fall in, a safety jacket was required because short exposure to the water could certainly kill you in a short bit of time.

At the Lac du Sauvage mine site, 1992. CBC radio did a documentary about us.

Here we are at a nearby shack with Claude our Quebec friend having coffee for that day's fishing trip as he told us of a good spot on our way out. Thanks see you later. Skinny and I were walking around holding on to the net and pack sack on a great spring day just off a Yellowknife river as we made our ways to shore. We carefully walked on the ice of the nearby lake, which had some open waters in the early spring. On our way back to shore for the truck ride home, with a good catch of the day, I had a view of a big pike. We stood there looking for movements to prove it was indeed a huge fish lying in the sun right along the edge of the river on a sunny, ice-melting day. "No that's not a fish," Skinny said. "Looks like a wooden stump." We doubled checked just in case it was one and just stuck our net in front as I scared the other end of him with my fishing stick in hand.

It took off under the ice taking Skinny's fish net out of hands in the blink of an eye. Yes, he nodded, it sure was a big pike, in comparison with the stories of others that they were truly out here. At least it sure proved that fish-tales might be real when people pulled their hands apart.

Celebrating the fish we caught that day, we headed back to town with a pit stop at the beer store followed by our discussion of why his first marriage did not work out the way he wanted it to. We sure had some fun times. We laughed about the champagne bottle I had shaken. Whoops – popped the cork off the bottle by accident and it flew and hit the minister in the edge of the neck with good timing. Just as he stated, "You may now kiss the bride," then off the wall it came over to the other side. The bridesmaid caught it in her hands with a smile, and said, "What a shot."

Yes, he had broken up with the wife in down the road, taken up with a wild lady from Newfoundland she sure was a party lover with quite a few laughs at get together for Ace and I. Here it was now 1990's and my good friend Ace was ready to leave the Yellowknife region like a few others do, for a change in life or just a break. We sat at my trailer at Trails End in Yellowknife on a sunny day, shooting the breeze and playing some crib, enjoying fun times and tall tales about fishing, along with beating others who were competing for a potential 29-hander. Yes I did get one, a while back. Here now we were getting an offer of two weeks in Whitehorse, with a ride and shared gas there and back. I had just finished doing yet another lengthy bush stretch and had a long shift driver on my cab car #8 by the name of Animal. No one messed with him, he was 6'8 and an A-1 driver well liked because he enjoyed long shifting. Yes indeed, sure a good time to get ready for a break fishing for some salmon off the Yukon River system, along with the great summer events. We talked about events in Whitehorse. Along with a busy summer every year, it is also part of the mining industry of Canada's storytelling. Anyone who had been there, participated with a tall tale.

Yes, I used my military chef/cook experience in the Yukon and drove a little hack out there too. Well I am in. Come-on, let's go. You can't beat this deal as the camp-style living is cheaply done. With good fishing every where I went along with sending some to friends.

After a few cold ones on this weekend, we decided to head off to the Yukon for a visit. We all packed some tents, gear, gas, and money for the two-day ride to Whitehorse Campground, with our good French buddy, Laval De Quebec. He had booked a two-tent spot for us at the well-known Roberts Service camp ground, aka the "world's meeting place," which was nicely near the downtown area of Whitehorse. Wow, just next door to the hippie yuppy bar – what are we waiting for? Yee haw! Let's go! The two-week camp-style event was so memorable that we made the news with our time spent there. We met a few others at the dance halls, and along at the Ear Lake a de facto nudist beach. It was best to have an eye open at all times... as the grizzlies are huge in that neck of the woods, yes I did meet a big one from a safe distance.

I had gone the year or two before to the same campground and best of all got to see my good friend Groomsy's parents, along with his awesome granny. This biker fellow had stopped over for a visit with motorbike alongside. "I got one just like that." Granny stated. The guy looked at us wondering if she was serious, and we all shook our heads in the yeah mode. He offered to take her for a ride on it. Gosh golly gee, I got to tell you. We got to see her ride a 610 Kawasaki round the camp highway, off the hill of the Yukon River entrance-way. With a big smile, she invited us to hop on with her for the ride, I got to tell you. Sure the most memorable family gathering ever.

With now the same wife and yes, two great sons and likely to be a grandfather good ol' Ace had sure pulled it off. He beat me at writing a book as he's already written his second. So here I am writing my first one. He has helped me a wee bit and we still meet now and again to chat about our tall tales of the fun times when we lived in northern regions of Canada driving taxi. He drove trucks along the way with a few other skills put to use.

We both met our loving wives through the worldwide, international dating service accredited since 1974, Cherry Blossom. Ace showed me the letter he had replied to. He was off to travel halfway around the world to meet someone. I took off out of the Yukon a day later to head back to work at yet another isolated, potential mine site of the future, with good fishing opportunities, along with airfare likely on Buffalo Joe Airways. Out there in the Great White North to make bread in more ways than one, using my military skills, along with the registry as a prospector for stocks and shares of the diamond industry. Sure why not? Drive a little hack in between. Alright let me know how it turns out for you Ace, sounds like a good trip.

Well we shook hands and off I went to that future mine site of today, hired by phone once again.

As a bonus, I had my training as a certified P.S.W. along with level-two first aid, which is needed for large groups working in isolation to increase better odds of, of course, minimizing deaths.

Although I had met my wife online, a little later in 2006, my good friend Ace had met his wife with a pen through the mail system while out of the Whitehorse. I got to read some of the letters just before he was heading out. She has a sister he told me when he got back, thanks for the info Ace.

Yes I was truly looking forward to seeing how he made out. Well, a little while later, he told me that she was on the way. "So stop over for a visit." All sounded pretty good as he described travelling and meeting folks with differences in language. "She is a Catholic," he said. All good, no worries he had pulled together a trailer for her arrival. Sure looked good.

Well, there in Whitehorse for a summer, a little over a year later, I got to meet his wife as a mother with a young child. They had a nice little trailer off the highway. As I had done a little stretch near my old army buddy out in Prince George, BC, I had packed the truck and off I went for the visit to finally meet that new mother, Beth with my good friend Ace "Dad" Cudney.

"Hello Dad. Well by gosh, sure worked out for you, eh? Sure enough." I looked around the online site myself and got to meet a nice Catholic lady from an awesome great family. I was looking to stop working in the bush.

After all the moving around for work and with travel adventures done, I decided it was time for me to settle down. Finally, one day online on the Cherry Blossom site I met a nice Filipina lady, who plucked on my heartstrings. Flew out to a beautiful Philippines Island to marry her and enjoy life together in a fun way. The trip was a quick three-week tour where I got to meet all her wonderful family, who welcomed me with enthusiasm. As my parents, friends and family could not make it there with me, we made a CD of our wedding. We brought it over to be watched with a bottle of champagne upon her arrival to Canada.

CHAPTER 11

DIAMOND TOOTH GERTIES

1982/ 83

Dawson City, 1982

Working in Dawson City and the Yukon, 1982

Winter in Yellowknife I was driving taxi, so now here it was winter of 1982 and I'd broken up with a girlfriend and moved to my friend Denyse Kutelsa's place as it was difficult to find a place in the Great White North of Yellowknife. So as she worked at the military section of Yellowknife at the time, I told her I had been in the military before and she said they are always looking. The following day I went and reapplied at their office for a position with naval radio communications. I was looking to reapply as naval radio communications officer and yes for the second time I passed, got taken in and worked with radio communications on the ship. They were going to send me back to Boot Camp in Cornwallis, NS and redo the twelve-week training, so here it was. On a spring day, I gave my two-week notice to my Yugoslavian taxi owner and told him I was rejoining the army. I thanked Denise for the info and a place to stay for a month, and off I went. So here I was at the CFB Edmonton. I was on a direct flight back to Boot Camp when all of a sudden as I stood there with my bag in hand I heard my name over the paging system, and that I had a call from Yellowknife. It was Denise. The lady I had broken up with was right by her side wanting to speak to me. As soon as we got to speak to one another she apologized for kicking me out of the apartment and offered for us to meet up in Whitehorse at a friend's place and start fresh. One of the military gentleman who had spoken to me about the phone call told me the plane was leaving in minutes and to say goodbye and hang up. I waved at him and said thank you but kept on talking to Theresa. Along with her apologies and her offer to making a fresh start, she said she'd pay for my plane ticket back to Whitehorse if I would be allowed to leave by the military. I could be charged for not showing up. As I continued to speak with Teresa on the phone, she was crying and once again asked me if I was willing to fly and meet up with her and not join the army. She said to tell them she was pregnant.

I had to sign off some documents to make my leave official. That's how I ended up in the Yukon. I took a taxi to a nearby motel and Theresa and I got a room and made up. The next day we started looking for a job. She got a job as a waitress in the hotel we were staying in. I looked for work and I got to meet other people from Yellowknife who were also on the hunt for jobs. The employment center helped me with my resume. Meanwhile, we were visiting downtown and getting familiar with the area. So finally, as the week went by, I found work at the Westmark Hotel where two dishwashers and the cook had recently quit their jobs. I was hired on the spot as sous-chef but I was asked to do the dishes first because the sink was full of them. Then later, I put the cooking skills I'd learned in the military to use. Since the job of the chef/ cook was seasonal, I was done too. I heard from a few miners in the area that I'd make bigger bucks feeding the workers in the mines. Having broken up with Teresa yet again, I flew back to Yellowknife in the fall.

As I returned to Yellowknife from the Yukon that fall, in preparation for winter, I did get myself an affordable shack in the Old Town just around the corner from Ragged Ass Road where I got to meet a good buddy Skinny/Ace that day.

I touched base with Skinny during our hacking work days, as we sat in one of the zones waiting for dispatch to call us on a trip from the Old Town part of Yellowknife to the back-up town zone. We were crossing our fingers for a twenty-dollar airport trip. The town was divided into zones and we were in the large parking area of this fitness club, waiting for our turns. All of a sudden a morning walker with the nickname Comminco stepped out of the Racquetball club /shower joint. He was a fellow I had gotten to know in YK and he told us someone had a shack to rent just up the road. Well sure enough, that day with the phone number and address of the shack rental, I offered him a cold beer in thanks after my shift. Sure enough, within the hour I got the shack from a long-time resident of Yellowknife with rules and regulations and the first and last months' rent. At the pub called the Wild Cat Cafe a little later, with my buddy Ron, the fellow who'd told me of the shack, we met another chap from Cobalt, Ontario, as we sat shooting the breeze about working here in the north. I got to tell my story about my Yukon adventures, as he was a diamond driller that came from a well-known mining community the past, which is now a retirement area because some folks never leave – that is where they stay to their last days.

This is where I came to hear about work in isolated regions for the big dollars. Of course, long, hard workdays are a part of it. Getting to be the cold days here now in the fall as we ran into each other on an early-morning, while chopping and hauling wood outdoors before it got to forty-below in Yellowknife. He lived right on Ragged Ass Road just around the corner from my fine, warm old shack. I had moved to the edge of Great Slave Lake with a view of the Northern Lights. We had both worked for Midwest Drilling in two different locations of the Great White North. That outfit's main office was in Winnipeg, Manitoba and they worked on making the best bid after the geologists' exploration samples start a mine.

As we spoke about mining, having been in the Yukon, I told him about the similarity in stories you hear. We went for a cruise later that week to the Yellowknife office to see when the next stretch was going to happen. The boss told us get ready because the ice was thick enough with a big contract in place for a two-drill crew.

That was just awesome. We got in early, and set up the six-tent camp for a couple of weeks ahead. Should the process be smooth with no problems, it often did an other 6 weeks.

Now, after a twelve-week bush stretch, we had done just about seventy miles past Yellowknife just off the Beaulieu River where a previous mine of the '50s that had existed before was just now being rechecked by a local group as part of the exploration crew. Often this was done at the start of the winter when the ice is okay for the drilling-discovery start of exploration, to next be moved to the on-land exposure for the finish of drilling, in search of minerals for the mining industry. Over time, working with the crews on the two twelve-hour shifts, you learn a lot about many things going on in this project. As we had been on a few runs he told me he'd read a book that stated only one-tenth of the gold in the Yukon had ever been found. Of my few past visits, he just told me that one day he was heading out there so if you're ever in the neighbourhood look me up and I'll get a shack out in that neck of the woods. "Alright," I said. "See you around. Have a good trip out there, eh?"

Jake's Corner, a nearby restaurant that I worked at, owned a cabin in Tagish, just on the other side of the river. With a turn off the Tagish road to BC northern, with a bit of flake gold in that area, bring your pan. The Yukon Territory borders British Columbia, Nunavut, NWT, and Alaska. It is known for its spectacular mountain scenery, along with awesome

Inland glaciers just off the mighty Yukon River.

The highway heads up to a few small communities, all the way to Inuvik, NWT. The Dempster Highway, which drives all the way to the Arctic Ocean, depends on tourism, some mining, and the legendary stories of days gone by, of gold-panning, sluice-boxes and open pit as one of the biggest Serpentine mines in the world.

We arrived in Whitehorse during the time of the winter carnival in February 1982. It was a mild winter and the locals told us that finding work would prove difficult. According to them, winter was not especially busy. This was a slow period and with a bad economy, but there might be jobs with accommodations included if you did not mind working out of town a ways. A lady who worked at the Chilkoot Hotel told us they had good winter rates with all the amenities. So we hit the employment centre, with little success, however. As we made our way to some of the local stores and pubs, my friend Theresa lucked out and got a job at a saloon, part-time. Well that was a start, so we headed to the hotels, and spent a month searching around with a car, since we had friends who had wheels.

I was working a busy summer season at the Sheffield Hotel, the old Travel Lodge in White Horse. My position was that of swing-shift cook, stocking up menu items for the next day, prep work for the salad bar, and covering shorthanded shifts. The hotel boasted a guaranteed seating of 300 people, mostly Americans funnelling in and off from bus tours for the Frantic Frolics show two a day. After about two months of that, the chef approached me with an offer to transfer to a smaller hotel the company owned, 350 miles north of Whitehorse Yukon. I had never been to Dawson City, Yukon so I jumped at the offer to relieve the chef up there. He was heading to hospital for surgery and would be laid up for a while. So I packed my bags and grabbed a ride with one of the tour buses. In 1982, the year I got there, you could still brown bag a liquor bottle on the street and buy booze seven days a week, and the Capital Bar in Whitehorse opened at 9:00 am sharp. Every year they have a re-enactment of a bushwhacker event.

My job was replacing Larry, the chef at the busy Gold Nugget Hotel which served bus tours and the local population. There were about 3,000 people in the summer. The hotel staff comprised four cooks and half a dozen waitresses. The hotel itself was a quaint set-up of about twenty rooms. The kitchen was small but efficient for the 80-seat restaurant dining lounge that was elegantly decorated in saloon style. The food service area had two large steam tables, salad bar, and a full menu with all-day breakfast and baked salmon fresh from the river. Open seven days a week to accommodate the busy season.

Muskeg welcomed the challenge. This was going to be fun; eight weeks of this, plus free lodging – another reason to go .Well wouldn't you know it, two months later I was still there.

The girl I was seeing at the time came up for the long weekend to check out the town. She had never been to Dawson either. I was there a couple of weeks and had started to get familiar with the town, so I took her on a tour. As we walked by one of the local pubs, this one dubbed the Snake Pit, music and the sounds of people carrying on and having a good time poured out into the street. The six-foot wide wooden sidewalks at front of the pub had some of the local canines just laying around. It was a little eerie We heard a ruckus from the Downtown pub across the road; a nice-looking pub with accordion-style stairs and swinging doors like *Gunsmoke* and the Old West cowboy movies. All of a sudden, this guy came stumbling out and landed at the bottom of the stairs, guitar to one side and hat on the ground beside him, followed by those famous words. "Don't come back here." My girl and I looked at each other and started to laugh, heh just like on TV. The man got up, and dusted himself off. His pride was hurt, that's about all. He smiled and said, "Your mother wears army boots and dresses you funny," to the doorman. Then he grabbed his things and with a bit of a power wobble made his way to the Snakepit, just across the road.

The downtown bar looked like an exciting place so we went in have a look around and sure enough it had the makings but there were only a dozen people sitting around. So we sat down with our backs against the wall. The waitress came, and we ordered a beer and checked out the surroundings. There was a stage where entertainers would perform later we were told. Shortly after we got our beers, this well-dressed gentleman with a beard and a captain's hat approached us and he introduced himself. "I am Captain Dick, a mighty river rat, from these here parts," he said with rough voice and a bit of an accent. He said he had been around for years and he had run boats and ships. He asked if he could sit down for a beer.

"Yeah sure," we said. "Why not?" So he pulled up a stump and sat between us with a big grin on his face followed by a smile, as he'd asked very politely! He told us he knew where to pan for some gold. Could we buy him one and he would tell us of about Dawson City's rich history, along with true stories of this mountainous Klondike region? The history of Dawson, eh! "Well Captain, ya get a beer," was my reply. "Give us the lowdown. Why did that fellow get kicked out earlier?" we asked.

"Ah well," he said laughing and replied that it was likely for bumming beer off the customers. We all started laughing and after about four beers, we ordered something to eat. Before the food came, the captain pulled out this gnarly-looking object from his pocket.

"Good golly, what is that?"

"My son my son," he said, "It's the infamous sour toe. Right, so how does it work?" He then told me the story about how a Frenchman had lost his toe in a mining accident and that he'd somehow ended up with it as a souvenir or won it at poker. It had then been pickled in vinegar or something. At any rate, for a small fee I could become a member of the Sour Toe Drinkers' Association of the Yukon. Since we had bought him a few beers, he tossed the toe in my beer glass and said, "Cheers, drink your beer and that's it," as he pulled a certificate from his pocket.

"Cheers," I said as the beer went down followed by the big to toe with the gnarly nail on the end.

"Oh ho ho,"" the Captain said as he shook my hand and gave me the certificate. "Welcome to the Sour Toe Drinkers' Club."

I did not think it was a real human toe at the time, however, I found out later that it was. OOOH my!!

We completed the night with walks around town and a ride across the river on the Yukon Lou, a small paddle-wheeler that ferried people to the campground a short distance away. There we ran into a gentleman who played guitar. "Hey, you're the fellow we saw at the bar earlier."

My friend, a few nice memories of Dawson City ...Too funny, small towns eh?

About a month later, I was at Diamond Tooth Gertie's the local pub that is a well-known gambling and poker place for Canadians and American tourists. There's a Can Can style show that is well worth the view. As I gambled one day, there was this local gentleman who had bet the pot on the last hand of poker that evening. He showed me his good hand but was short 200 dollars for that bet. He offered up half the pot or the ownership of his car. Fair enough. "Does it work for you?" I asked.

"Hell yeah," he replied.

I shook hands, accepted, and gave him the money. We waited for the dealer's outcome to tell the tale. Ah shucks.

He did not win and we took the car out to Eagle Plains for a test drive. Hell, we saw a huge grizzly. Yes, had to hitchhike back – took the first ride. About a week later, I stopped for a visit at the place where the man had sold me the car, and knocked on the door. His wife told me he was sleeping and to go and wake him up. So I did, just to let him know he could have his car back because it had only made it halfway. When I entered the room, the man was sound asleep with one leg sticking out from under the blankets. The big toe off his left foot was missing.

Suffering succotash, I thought. *Maybe he lost in to the captain in a poker game.* Oh no, later he told me he'd lost it in a mining accident.

So off to Yellowknife I return, back in the fall of 1982 right at the YK Inn with a place to live, or so I thought. Sadly, due to too much playing with renovations and its expansion, living quarters had disappeared. However, the position as sous chef was always in big demand. Back at it for me. I made a visit to my bro and wife at new place they called the Dog House. Well this was a party place of its own and I was finally trusted to be put in charge of rent with tenants. Having gratefully stayed there for a month, I later got my first affordable shack in the Old Town. It was like the days back, with a one-room 12 x 16, water delivery once a week, a lantern for back-up, and an old style wood stove on the main drag, just off Ragged Ass Road for inexpensive rent and a fitness club a short walk away for showers. You cannot ask for more. I still drove hack part-time, with time to compete in crib with my Skinny Ace and a few others with names used as precautionary measures, should you get in trouble with an evening out at the Strange Range bar, as many did to lower the stress level of forty-below winters. Soon I was looking to do another bush stretch for well-earned dollar to pay the bills

With no car in hand to get me around, I was taking a few taxis at forty-below chatting and getting advice from a few other cabbies on getting my class F/4. Well here we go, I started to drive taxi part-time back in 1981, along with potential day offers by other taxi owners in the future. In a different line of work, I got to meet my long-time friend Ace Cudney, who was a cabbie at night/

trucker by day, and buddies with Alex Debogorsky. Code-name Skinny/Ace, my good friend to this day. As he beat me in writing a book, he helped me to write mine as we traveled in many places, sometimes with the same group or with others as time went by. Through this process we did indeed get to meet many folks and their families that I am still in contact with to this day. And now with today's online access to communication via email, Facebook, and texting, it is all so easy. Sure glad to have learned to type. He and I had dated few ladies, he then, at thirty-eight years of age got married to a lovely lady from the Philippines. They now have two boys who live not far away from each other. My loving wife is also from the Philippines. We have gotten together to talk about the old days spent in the Knife. Like many who had the inspiration to come up north and be the hard workers, with their kids, they continue to be friends of the family. I got help in reminders for writing this book. And Ace gave me help in going through an Acquired Brian injury as he is a survivor himself from a crash back in the early days, living in Ontario.

Yes I have made mistakes, and learned the hard way while in public service about the rules to abide by.

As City Cabs of Yellowknife was bought and sold to individual owners more than once over a time, it had a good reputation with drivers from various parts of the world as a second job. Some of them were miners, and others were full-time locals. Working extra hours for big dollars, especially in the winter was in big demand. In the early '80s in the Knife, I still worked days at the hotels, then started to drive an evening taxi shift to make a few extra bucks. Busy weekends providing transportation for a large part of the growing community of around 5,000 at that time, including many local Dene as well Inuit folks flying in for a break from other northern communities. Especially during winters of forty-below, cabs were indeed in big demand to this day, and many drive sixteen hours.

Along with a few other drivers, we played crib or backgammon if it was slow, or after the shift we shot for the 29-hander for peg motion with that deck of cards on the cribbage board. After shift, our group had a cold beer at someone's home or we sometimes played at the bars on slow days where Sam from the Range had a few tables in the shape of a crib board and the availability of new cards. Not a poker player myself, I met a few that played and won as well as some that sadly did not. They destroyed their lives by gambling too much, either in the stock market or the many other avenues available. A little goes a long way. There were a few ladies who challenged us now and again. If you got broke Sam himself would offer you a flip of coin for double or nothing to pay back on your next visit.

Ace and I talked about writing a book for years, but that son of a gun beat me – he already wrote two. Ron is his real name and he had played poker once against a cabbie to win half of his car, to then later sell it back to him or something. Trucking opportunities hauled him out of the Knife as he had decided to move to Whitehorse in the early 1990s. We still kept in contact by phone and now by email as he helped me write my book. Did a few bush runs in the Yukon, had a chip wagon on the go for a bit, so each time upon my arrival there was a call for a beer with a game of crib to catch up on the latest Whitehorse chats.

We are still friends to this day. One of his sons plays hockey and the other plays a trombone. They both date girls that will make their parents proud in their future as grandparents.

CHAPTER 12

THE NORTHERN MINING INDUSTRY

Safety hours, shares, news and movie clips

Feel free to mention me and, while it might be a shot in the dark, to also contact Phil about possibly hunting down that documentary. January 01 2015 on e-mail.

Well known Journalist Mr. David Miller has 25 years in broadcasting at CBC with an achievement award from the Canadian Institute of Mining. Holding 26 awards in radio broadcasting MÂS'KÉG MIKE was sure glad to be part of this as he flew at our exploration site to spend a day with us as a snow storm was on the way he almost almost stayed 2 days. Lac du Sauvage 195 miles north of Yellowknife NT, connects to Lac du Gras off the Copper Mine River system where existing mines of today have many hard workers from all parts of the world. Yes this is the one I have help start in many aspects along with providing public transportation between bush runs to participate at local events in Yellowknife NT.

Annual Caribou Carnival of Yellowknife NT with dog mushing events, skidoo races to name a few. I got to ride the sleigh dogs on the Great Slave Lake with a few of the best racers in Yellowknife NT. Summer event with song festivals with tourist and famous folks that come to a fish experience in 24 hours of sunlight.

Started driving Taxi Yellowknife in 198, later in the Yukon & ON. Owner operator of many taxis in Yellowknife with City Cabs 93, a company I helped start in 1992 at Tim Hortons YK. Stock market manipulation to a profit is part of our life no dought about it. Of my half a dozen shares I owned like many miners from all parts of the world did for many years. My military skill enabled me to teach 2 cooking courses one to the Dene the other to Inuit students. Hand build garages of all size along with plywood boats with a few birch bark canoes I helped with. Small house boat out of 45 gallon barrels sold to a local to live on. I have fished on all 3 coasts of Canada in many isolated regions of the mining industry in Canada. Helped start a few mines, one of copper in the Yukon the other with Diamonds as registered Prospector. Half a dozen movie clips as proof of my achievement story-telling festival in Whitehorse & Dawson City Yukon.

Mines an exploration sites in Canada located near lakes and rivers I worked at some near towns others at new sites with Geologist / Drill crews /Miners/ Plane and helicopters Pilots. All looking for, Gold -Silver - Uranium-Copper- kimberlites / Diamonds.

I mostly fed crews of small and large numbers living in canvas tents with a priority as spring bear watch. My learned skill as a PSW / First-aider level 1- 2, for extra bonus pay that was required to be hired for work with many of these outfitters. As a registered prospector helped staked claims for two of these mines in Canada. Worked for RTL, ice road trucking of Airport Road of Yellowknife NT on various pit stops between to the mining areas. Core splitting for a geologist part of my six week training with a one month as a drillers helper sure brings a day's work to reality of an others workday. Drove taxi during Giant mines bitter strike.

Fished and hunted on both ends of the Copper-mine River, part of a radio and movie clip in 1992, Diavik mines. Midwest Drilling pay checks with over 10,000 safety hours in them there hills various parts of the north. Taught a cooking course to YK Dene students, then an Inuit in Kuglukyuk.

Company names with the over thirty sites MÂS'KÉG MIKE worked at 1981 till 2011.

Quyta Lake **Con mine** outcrop, **Giant Mine** outcrop Pointe Lake 1991, **CBA** Beaulieu River mine,Lac du Sauvage,(movie -radio documentaries) Staked claims for Diavik Mines of today. Salamita /Tundra Ruth mine sections, **Midwest Drilling YK**-Thompson / Lundmark off North Slave, **Echo Bay/ Lupin Mine** in town & at site ice road truck loader & Hercules C-130, **BBT Expediting** of YK as a labourer hauling full core boxes on site at George Lake NT. **Midwest Drilling YK S**lemon /Courageous Lake, NT for cleanup, **Colomac mine** as a drillers helper, **Fortune Minerals**- Colville Lake near Fort Ray / Becheko NT. **Discovery Mining Services** at their original site of Discovery Mines & Long Leg Lake, Chicken Lake for a few good stretches.

De Beers of Canada, I worked around Baker Lake to an easterly direction, Snare Lake river system canoe route Lower -Upper Carp / Snap Lake for Gahtcho Kue (Kennedy Lake) Fingers lake area. In 2000 I taught a cooking course in Kugluktuk Nunavut.

Worked at the **Marshlake Marina** just south of Whitehorse right by Jake's Corner pub with great view. Prospecting as part of the copper belt south of Whitehorse had my name used at mine site sale 1990's. Mr. Carter from the Cobalt ON mining community was a diamond Driller co-worker I did this with. Haggled at a poker table Diamond Tooth Guerty's of Dawson City won an old car, later traded another vehicle for a cabin. Owned that cabin in

Taguish off Yukon River, 40 km south of Whitehorse near Skagway Alaska border.

Aurum Resources Ketsa Mine, 70 km. south of the Ross River, Yukon 1,700 fts drilling project. Panned gold in Dawson City as I worked at the **Eldorado Hotel – both Westmark** Whitehorse Yukon. Whitehorse two weeks start toured the Top of the World Highway.

Domco Foods: 1986 /87/88 Pickel Lake ON, Red Lake ON, Geralton ON located near the old Diefenbaker Hwy near, Nakina, then at Foleyet mine site very near Timmins ON. Train ride to Moosenee ON. Over all 2 years safety hours at a dozen mine sites, fire fighters crew, Junior Forest ranger of Ontario. Stop over at Empress Mine site, 12 weeks work stretches as part of agreements for travel time and bonus pay for cost control safety first .

I taught a cooking course in Yellowknife, NT in 2000

Upon my arrival back in Yellowknife, I drove a little bit of taxi with the seed of hope in the back of my mind about working in the mines as a cook. In the spring of '1983 in Quyta Lake. I did my first in-camp session as a replacement worker for my friend Robin's wife who was on maternity leave, she worked for a rehabilitation camp. This was my first three weeks in the bush, living in a canvas tent with all the food, and we had a generator as our main source of electricity. I worked with Aboriginal people who were from the jail's early release program. An Aboriginal elder trained us how to live off the land even though our basic commodities were store-bought. After three weeks, with the session coming to an end, I was leaving camp with a take home of fish and caribou meat. My first bush run sure impressed me. It also gave me the chance to catch up with Robin and meet his first son, we both lived in Trails End Trailer Park.After the winter of '83 driving taxi full time, a catering company Crawley & McCracken Co. Ltd. Hired me on the phone, position turned out to be as a Bull cook out at the Salamita Mine site. I was the cook's right hand and did the odd jobs in camp peel potatoes, wash dishes, fetch water and keep the camp clean make beds. I put in a full four weeks until they got a replacement worker for the job I was doing, because that was not my original position agreed upon. I had been on extra duties while in the military, it sure gave me fond memories. Even to this day the Canadian Armed Forces do a winter inspection with a volunteer force of the Canadian Rangers in the North Warning System I have seen them from the air on one occasion.

Shortly thereafter, I headed back to town and got another position as a labourer, hauling core boxes and re-piling them in order of past samples, which I did for four weeks. Getting paid $100 a day. We used two emptied 45-gallon barrels with an improvised wheel based wagon with drill rods as axles and 2x4's nailed together to hold them in place while we hauled the heavy core across the tundra. It was better than carrying them on our backs.

Here are some of my tips for planning your own survival in the Great White North: Survival Tips of Isolation learned with some new ideas of *MÂS'KÉG MIKE* and others. These skills were part of my military training, pass on to simply survive another day. To work in the isolated mining industry MNR info classes include proper rifle use, and awareness of bears and all other wild game. Firefighters of Yellowknife have a one day class to properly use the ABC's of all fire extinguishers.

I trained as a Personal Support Worker combined with my Level Two First-aider courses, a dozen times for isolation. Security skills as a Commissionaire combined safe driving in the public transportation service of driving taxi, which has brought my level of awareness to personal care up a notch or two. Military crews, still keep the isolation-survival training to this day, I know first-hand, having spoken with them just before they head out to do so. Now being part of the labour code in Canada, WHMIS education ascertains the understanding that covers important aspects of safety over-all. One good example: one year at fifty-two below zero, out in the middle of nowhere-land, with no heat for a whole three days, we had our diesel oil freeze up in our workplace. We were out in isolation close to the North Pole region for a winter project of diamond exploration with a geologist and a basic crew of less than ten. There were three tent quarters for us all to make use of the five-star sleeping bags, where the cold can freeze your body parts easily, as it had already done to many northerners. Keeping an eye out on each other is a reminder that we are not made out of steel and to use your hands to warm up away from the wind,- Should there be an emergency heaven forbid, working together is essential.Being prepared for emergencies starts with a plan It's just like camping and a first-aid kit is number one priority. How long would it take to empty your food stores? Less than a week? Keep dry good supplies. Pick up things at garage sales as you build your survival kit to be ready- tarps, nets, and camping stuff. Show your kids some skills. Make it fun use five gallon plastic pails with lids for storage. Wrap gear so it can easily be carried and mark it as **Survival Gear (30 lbs).** This is only an example - Any basic setup where minutes count is a worthwhile project to start. Double up on a carrying bag with straps and make a list of important items. Keep weights at a reasonable carrying size- Styro-foam is as good a floating device as wood is. Fuel barrels or plastic items sometimes left or caught by the wind can be found to use.

Compass - axe - shovel - saw -2 knives- flint-stone and a file, fire extinguisher.

Dry goods of food items like bars, rice, fruit, vegetables and meat can items

Sewing kit scissors – cotton wad - good whistle (snare wire- rope- tape)

Full First aid book and kit - Tarp - Lighters and waterproof matches, Kleenex

Canvas tent mat, sleeping bag packed in order, a toboggan.

Pots and pans, cups and utensils - for 2 sets of rain gear (garbage bin and bags)

Flashlight -flares– candles – dry paper tissue – nails -air horn

Dry/smoked food items plastic holders (rice, pasta, flour, tea, canned goods)

Warm clothing - fishing gear- sling shot – crank-pellet gun for practice.

Wood kindling even if it's just enough for one day, kept dry in a plastic bag

Cell phone – binoculars – sunglasses – grill - bear mace- flares

Skis and poles- walking stick – crutches - nets or string -

Magnifying glass – survival pamphlet – string to make nets

Practice drying meat –veggies – pickling with salt, aluminium foil

Grow something. any germinate seeds – compost- clone plants

Learn to tie knots - weaving might come in handy to make nets

Ice hogger spare blades with chisel attached to a long line so you don't lose it

Two poles with a strapped jacket or tarp can be ideal for carrying, make a pole yoke for the shoulders, or the traditional style, neon marking, crank or solar powered flashlight.

Knot tying /weave or make nets traps using materials at hand or string /cloth anything

In the north you have fresh cranberries, blueberries, raspberries – root items

Moss is good for many uses - spruce bows or wood spears or canes.

Knowing the ph in clay - digging holes for growing soil - leave a few marked area for burning later as wolves bears coyote to name a few will make friends with you

Winter time fish abundance can be used to make a slay like the Inuit folks do then to eat later I was shown how being part of the sleigh made more efficient.

Frozen together to create a pole length of 6 feet then wrapped with a tarp or canvas material for shape of a sleigh is one of many like well-known whale bone wrapped with fat to make a ball.

Build a small boat in less than a day - you need 3/8 or ½ in plywood, or whatever

Saw - hammer or pencil - battery powered drill - a measuring device

Two 4 x 8 sheets of plywood. Use 1 to make a V with a one-foot nose

In the north you have cranberries, blue, berries, and raspberries – root items

Moss is good for many uses - spruce bows or wood spears or canes

Knowing the ph in clay digging holes for growing soil -leave a few marked area for burning later

Winter time fish abundance can be used to make a slay like the Inuit folks do then to eat later

Dozen good size fish, a few eaten keeping the skin and bones to jam together with warm water. Frozen together to create 2 skis of six foot in length, by bunching the fish together head to tail with with water turning to ice, circling the skin of the fish around it to secure as one. Then connect it side to side with the fish tale poke holes as needed use skin to wrap, ice to secure.

Build a small boat in one day – you need 3/8 or 1/2 inch plywood, whatever you got.

Saw - hammer or pencil- hand /battery powered drill-a measuring device

Two 4 x 8 sheets of plywood. Use 1 to make a V with a one - foot nose

Six 2x2 eight feet long to fasten all the joints together

From the 4 foot end take 18-inch wedge from the 4"middle of sheet

Cut 1 sheet into 3 even length slices, approximately 16 inches

1 quart of tar to seal all seams or gather spruce gum mix with ashes

1 lb of ring nails or drywall screws- glue is good to bind inside

Paddles made from two hard-wood trees and 2 wedge attachments

Attach a rudder mast with a piece tarp to catch wind

A piece or rope and an anchor rock or chunk of metal

Doubles as shelter-the fish are in the deep spots easy to catch

Use empty barrel as a stabilizer or Styrofoam or Javex bottle -air bags of any kind.

Use sand bag rock as anchor, winds can be strong at times to knock it down in the snow.

A raft with dozen poles of 8 feet or so tied with ropes duct tape or whatever you-have in hand.

Radio Communication at that site /fire hives, empty fuel barrels/ Pyramid/marked landing area/spray paint – orange plastic bags – any wood items available that insure a good grip.

Compass is not as efficient due to the magnetic dip closer to the pole location. Double check before by hand, as a verbal confirmation of a magnetic compass to assure accuracy directional gyro in the air can partially cause a difference.

Mark it down; double check locations, with a second opinion then you'll have a better idea for northern regions or low level areas. When pilots come by confirm their flight directions for landings; Wild game animals have tracks along with their inner locations on the south side.

Should you be stuck somewhere you need a radio and transmitter with a spare one that works. High quality hand held radio transmitters. Most are powered by battery and require an antenna set-up. Go to highest point on a nice day upon arrival with a backup of modern technologies. Sturdy antenna to last its use using heavy rocks, bags of sand or digging a hole around it first.

Phonetic Alphabet (Internet)

A Alfa I India Q Quebec Y Yankee

B Bravo J Juliet R Romeo Z Zulu

C Charlie K Kilo S Sierra

D Delta L Lima T Tango

E Echo M Mike U Uniform

F Fox Trot N November V Victor

G Golf O Oscar W Whiskey

H Hotel P Papa X Xray

Create a landing strip or spot for helicopter. Looking for flat open spot, or a lake. Shovel the ice some.

Ice hogger to check for thickness of the ice, of the distance with spare blades for thick ice. Landing strip 1000 feet in length marked at every 50 ft with handy items.

Mark with bright colours (garbage bags) 50 feet square diameter for helicopter and air-planes passing by to see from the air surface. Ice fog is a very true subject, freezes parts aerial devices.

When giving weather reports refer to sky as ceiling low or high on the radio 5 by 5 means A-1. Giving awareness that any lower means view and winds make it difficult to land at this time. Spray paint as a marker. Small pine trees if available, rocks, or a stack of branches or ashes from the fire left in an obvious site.

Morse code (series of dots and dashes taping or with flash light)

A . - B - ...C -.-. D -. . E . F..-. F ..-. G- -. H .... I . .

J . - - -K -.- L.-.. M - - N -. O - - - P .- -. Q --.- R .-. S ...

T - U..- V...- W .- - X - .. - Y - .- - Z - - ..

1 .- - - 2 .. - - -3...- - 4....- 5 .....

6 -....7- - ... 8 - - -.. 9 - - - - 0 - - - - -

In the northern area

Access to this info marked in an obvious place, with a grid-lined map, marked with your exact location to start with then the area you came from. Use a compass to be careful about declination, as the sun rising in the east is minimized in the winter time large water-ways flow from north to south as makers.

A survival tip is using signs and basic sign language as sometimes others do not speak English very well. Let people know beforehand where you are going as part of a smart way to plan.

Set up fires to alert rescuers to your whereabouts - know fire safety not too close to the forest area to cause a worst case scenario. Keep water and branches handy as a preventative measure, there could be someone living nearby. Mark your area alphabetically with wood rocks or material at hand, so it can be easily read from the air HELP, Use red flags, coloured garbage bags or clothing items tied to something like a stick.

Make the shape of an arrow on the ground pointing to your location near a stack of wood ready to be lit.

The potential of all hazardous chemicals to cause a fire is part of safety in all workplaces.

Note: Remove either fuel,- oxygen, or heat on windy day, to a slow or end a fire, ASAP

Class A Green Triangle **Implement corrective action**

One of three must be removed Fuel /Oxygen /Heat to extinguish fire

Class B Red Square 1st to notice fire-**Yell Fire Fire Fire**

Trigger alarm - Notify fire Department (Alarm)

Class C Blue circle Supervisor initiates report notes

FLASH POINT IS 200 % F

Class D Yellow-star Spills dealt with accordingly

Knot tying is very important and worth learning with the basics, as old school techniques are still usable.

Survival days of our past still apply at any given time, having some idea of what to do while stuck out, in the middle of nowhere.

That is what the Metis sash design was made from, by knitting together with your fingers in various ways just like a fishing net- It is not just only a gal thing and is worth learning right from the start.

Isolated Use and Storing Lake Water -summer versus winter

Having come close a few times, feeding the crew in isolation, I would just hold off on ordering more than needed to avoid having to reload it again on its flight out. It sure made a cost difference.

Communicate before your order with local distributors to prevent mistakes of over ordering goods.

Water is a valuable commodity, and consumption is always a priority. In the winter, melt snow or ice. If water is scarce look for icicles. Usually you will find a deep spot in the middle of the lake or near moving water. Use only if it has been boiled first or has been run through an environmental kit. In getting water from a lake, have an experienced person show you how to work hand pump. Water pump 101- Get some training. First prime properly, then drain for storage. Use properly marked fuel and oil, and never leave water in the reservoir during winter.

Keep a wet rag or fire extinguisher handy, not to close to shore and stabilized on a platform anchored to rock area. End of the hose should have a screen or filter, Use big garbage pails and rinse with bleach once a week. Keep water jugs filled. Winter time you have to find the deep spots in the lake. Make sure you are familiar with the safe use of ice picks and chisels (rope on the end) huggers (extra blades) chainsaws.

One hour before everyone else rise and shine - usually about five-bells

Have an eye for wild game touring around. Winter is less, however, summer is more so.

You serve the big breakfast with a cheerful attitude and some easy listening music. It's important to be a good listener. First thing in the morning, get the show on the road and put the coffee on for the early birds. If there is power, have a stove-top pot of coffee prepared the night before, or instant to be on the safe side.

Have a spot set up for workers to make lunches. Some cooks like to have offers of all morning items every morning, while others do eggs one morning, pancakes another or French toast and boiled or poached. Because of the two shifts, many workers enjoy a hearty breakfast of beans, potatoes, bacon, sausages, or leftover steak from yesterday's food. Most workers like a routine. You get used to it and blend it into yours.

Tips:Use Saran Wrap or foil on paper plates or regular plates to save water

- #1 Safety in all aspects. Isolation can be life threatening itself.
- First aid is to cover all aspects. A flight out is the next step if too serious
- Hours along, beverage request - favourite cereals, snacks, and desserts daily
- Breakfast, if you can manage it, leave it open to individual choices
- Lunch -soup and sandwich, desserts, all the fixing or hot meal for 25 men only

- Supper - two choices every night, rice or pasta dishes, steak and-spuds once a week

- Walk up to stove or counter for the pickup of your plate with food choices

If someone got injured or needed to go to town, the chopper was our emergency vehicle, especially with fifty folks working in that area, as we were about 150 miles away from the nearest hospital. Mr. Shapiro was our experienced Japanese pilot. He was familiar with the area, and a good sense of humour. Shapiro was trusted and well liked for his chopper skill, in loading drill equipment from one site to the next. At no extra charge I was often the hook up man at our main site, and I got a good many free rides with an awesome view of the area for pictures and good fishing spots.

The time frame for a plane trip out to an ambulance then hospital could be four hours cut in half. Now and then the need to go to town would arise and certain people with a flexible schedule would be able to fly to town with a list of personal items to pick up for the whole camp. Flying back the next day with the incoming flights, all had superb views with an idea of the lakes and rivers for the trip back.

Legally, all injuries are all to be reported. Disposal of all waste if it is part of at the mine's end. For example "Steps of or processing "Should wearing safety gear be a factor of a confined workplace, being immediately reported by phone paper work online as a precautionary measure.

Chemical exposure or lack of oxygen example carbon monoxide due to a plugged up mask or improperly fitting you while at an operating rock crushing plant /strikes to name a few, were some possible dangers and were made part of our safety process so they didn't happen again.

The Natural Laws of Healthful Living
By Carlson Wade with a forward by H. W. Holderby, M.D.

North Bay librarian's info.0.1 Canada, General contact -CCOHS -Hamilton ON. 905 572 2981

Canadian Center, Occupational Health and Safety 1 (800) 668-4284 (Toll-free Canada and US)

International Labour Office (ILO) By telephone at: (905) 572-2981 or 1 (800) 668-4284 -www.ccoh.ca Workplace facts of Canadian mining up to date, used in the mining areas of isolation for safety. Training on site for all aspects of the work places- PPE, Personal Protective Equipment and note pad rule - Stop, Take 5, Think, Recognize, Assess, Control, Keep safety as the First in all Tasks!

WHMIS classification: A - Compressed Gas; B1 - Flammable Gas; D1A - Very Toxic; D2A - Very Toxic (Carcinogenicity; Mutagenicity; Reproductive toxicity); E - Corrosive; F - Dangerously Reactive.

CASRegistryNo.75-21-8 **Other Names:** EO, ETO, 1,2-Epoxy-ethane

Main Uses: Used to manufacture other chemicals, to sterilize medical devices, and as a fumigant. **Appearance:** Colourless-gas. **Odour:** Sweet

Canadian TDG: UN1040

Occupational Health and Safety Act

ONTARIO REGULATION 490/09

DESIGNATED SUBSTANCES

Consolidation Period: From January 1, 2013 to the e-Laws currency date.

Last amendment: O. Reg. 148/12.

Ethylene oxide

This Regulation applies, with respect to ethylene oxide, to every employer and worker at a workplace where ethylene oxide is present. O. Reg. 490/09, s. 8.

Ethylene oxide itself is a very hazardous substance: at room temperature it is a flammable, carcinogenic, mutagenic, irritating, and anaesthetic gas with a misleadingly pleasant aroma.

The chemical reactivity that is responsible for many of ethylene oxide's hazards has also made it a key industrial chemical. Although too dangerous for direct household use and generally unfamiliar to consumers, ethylene oxide is used industrially for making many consumer products as well as non-consumer chemicals and intermediates. Ethylene oxide causes acute poisoning, accompanied by the following symptoms: slight heartbeat, muscle twitching, flushing, headache, diminished hearing, acidosis, vomiting, dizziness, transient loss of consciousness and a (sweet smell) taste in the mouth. Acute intoxication is accompanied by a strong throbbing headache, dizziness, difficulty in speech and walking, sleep disturbance, pain in the legs, weakness, stiffness, sweating, increased muscular irritability, transient spasm of retinal vessels, enlargement of the liver and suppression of its antitoxin functions. Ethylene oxide easily penetrates through the clothing and footwear, causing skin irritation and dermatitis with the formation of blisters, fever and leukocytosis.

CHAPTER 13
SHEEPSKIN

Yellowknife to Canadore College

Canadore called our house looking for you to take the Meat Cutter's course that you had applied for on your last visit to North Bay, so will you be taking it let me know. Well, paying attention to my mom's advice as to what should be done, driving hack did not give me any official work weeks, so I had to use an adjustable approach for the EI. Weeks had to take place, oh well. Here now in 1985, I slowly depart from Yellowknife.

At least I had a few weeks under my belt for this year at the work place in isolated regions with a few bucks in my pocket. I prayed for that to fall in place. So I told my friend Ace we had time for a last crib game till next time. Being like brothers, we shook hands of course and had a cold beer in competition.

Winter was nearly over, with the crossing falling in, and the lock-down mode for departure. The time for leaving the Great White North was now, or there would be no traffic for six to eight weeks. There were reminders about this timing on the radio. Truckers could take a day by day offer, or you're out of luck. Or you could just get stuck in the Knife unless you flew out or walked with a goofy set of boots on the ice road.

The ice road was to be used to cross over near Fort Providence, land of the Dene/ Métis Band community, just across the Hay River, NT area, before it all melted away. Where the bridge had not been yet built, is today known as (DBDC). That said, leaving required transportation by car to make it past the Aurora Borealis for the trip down on the North Highway to Edmonton. Then largely from A to B without any breakdown to the home plate of my home town, North Bay, bo'y. Sure felt good as I had packed a few bags for light travel just in case you get rammed by a bad driver, a Buffalo, or did not make it to the river crossing in time, due to the melting weather of the twenty-four hours of sunlight on the way. I'll miss it, for sure coming back to YK.

Just after driving cab for a cash wage during the winter celebration in the evenings, now the very well-known and liked Caribou Carnival of Yellowknife sure gave you bragging rights of the northern region.

So now with my 1970s Dodge Coronet all set, which initially had taken me to the Knife, once again, here we go for the trip back. Sadly, at that time of the year not too many are traveling the road. It's still a little cold and melting farther down.

With ferry departure time announced on the radio and TV to prevent your waiting, time your arrival to the amazing spring view of the Mackenzie River, till the Ferry Crossing hops in. Nowadays, thankfully, the new bridge is finally built. The Northerners know of this pragmatic moment.

As we all get to find out that the *spring* breakup crossing is part of the season; short and sweet, along with the hazardous floating or jammed ice stacked not far away, sometimes there's a sound like an explosion as it opens up. Yes indeed, I heard it once while waiting at the crossing for the Merv Hardy Ferry ship.

No worries on a sunny day, to be first one loading at 6:00 am departure to cross the Mackenzie River, an about twenty minute ride with an marvellous view of Fort Providence on it's bank that is now called Deh Cho NT that is finally connected to the rest of Canada.

So off I head to southern Alberta region with a few pit stops in between. Ontario is said to be a four to five-day drive, depending on a few road issues. Should there be any, use your five-star driving ability, good tunes, and short sleep in the car. Ya sure, why not? With a sleeping bag all was good. So far so good; over 400 miles down the road on a wonderful day. Having a few out of province friends' contact numbers in planning ahead is worth it.

Well here I was, twenty miles away from High Level AB, one of the first fur trappers of the highland area of the 1770s. They were just a little south of there from my mother's side's family name. There is a Foisy Ville in AB. Yes, met a few folks whose family name is Foisy even in Yellowknife, as they have an annual gathering out in Edmonton telling you the history of where the name comes from; France/Quebec eh? There is still wild game to watch out for. Being careful while driving is just respectfully done, by slowing down.

Anyhow, at High Level AB, was a good, French hunting buddy of mine with an invite to go fishing at any time. Married to a lovely Aboriginal lady, and had four boys who liked to hunt moose, fish, and trap with dad. They always invited me to stop over at any time on my passing through, with a reminder; oh yeah don't forget a cold beer as part of the stopover cost, with a little bit of BBQ stuff for his loving wife to treat the boys. With smile in mind, though, looking forward to just spending the night or two for a break, and listening to some tunes as we will catch up with the news about the boys' family life out in this particular neck of the woods. Yes indeed, there I get to gas up at the nearby First Nation outlet.

My friend was originally from a northern Quebec area when his father came up to Yellowknife to work in the mining industry. That is where we met; at the Trail's End Trailer park back in 1979. Yes with the fun times, we toured the Gold Range Bar of Yellowknife as both celebrated our twenty-first birthday in the same month. He met the love of his life and now was a responsible father. With his offer to visit in place, I was sure looking forward to the stopover.

With thought in mind for a fun visit coming underway, all of sudden the car stops running, just after I passed the 20-km sign on the road, which shows how close to town you are. One can be there in fifteen minutes with a little cheating. So as I pulled over on the side of the road, a little steam was coming out of the hood. I slowly popped it open to check it out, just in case it was more than that. While it slowly cooled off, I stood there with my water bottle for a backup. A slight leak is a little scary, and I was hoping that with a cooling-off, the car would run me all the way to town.

An hour or so later I topped up the radiator with water and started up the Dodge of mine. Oh boy, thank God. Pray to make it at least to the community of High Level AB.

Making it into town at the first one on the right, gas station garage with a B and B along the side, I pulled up for the mechanic, a fellow having a coffee as he smiled and asked if there were car problems.

Nodding, oh yeah, good day. With the so so description of a possible leak yet. "Well when is there time to check it out?" or for me to fix it with a description of what happened.

I hopped in and drove around to the pump to fill it up with gas. Let it run a little so I can have an idea.

I told him about the water leak the car had.

He just walked up and cracked the hood as I gassed up. He said that the radiator was leaking and also could be plugged up by the looks of it. "Alright," he said. "Leave the car in the parking lot and come back after lunch and we will let you know, See you later."

Hoped I had enough money for the gas run with a little to spare on this repair as I headed back home, hoping mostly to get there. Trust and care in a vehicle is also very important especially when you're hoping it is not serious. First of all, I was just shooting for the first thousand miles to the Edmonton area. That was sure a big part of the driving trip, all the way back to Ontario. There were flights if need be, or bus fare, however, relatives and friends were there to help out too. At least I knew a family in this community. It sure was worth the stopover for a visit. With a good night's sleep before leaving for the next 2000 miles ahead. Making it back home safely was what it's all about. If the car gets fixed for a good price that I can afford.

Walking on over to this pub across the road, I would just leave my car parked there for at least one night as it was going to be a break by sounds of it, just waiting for that news. I stood at the counter of the pub after my morning brunch at the restaurant section, thinking about just getting a beer. Holy smokes, there stands a good fellow, a mining co-worker, a Frenchman, Mr. Michaud that I knew me from Yellowknife. What a surprise.

"So how are things MÂS'KÉG Mike?"

"What the hell brings you out here in the Level all the way from the Knife?"

We laughed and shook hands at the counter.

"Well. It's 11:00 am we just opened up," he said. "Want a cold beer? You can tell me what happened."

We sat at one of the tables and he told me about finishing an inventory. He was the manager at this bar and he told his co-worker he was good for the day, so talk to you later. I was impressed and sure thankful that I know someone else in this community. Cheers with a cold Canadian beer and blabbing it up to tell him what brought me here. He started to laugh.

"So what brought you out here, Jack?" I asked.

"Well, I had stopped for a visit to our Yellowknife buddy like just like you're doing, when I was offered the job managing this bar. So here I am. What do you think?"

Right on, well a place to crash is what I was looking for on the way to heading back home. Now with car repairs on the way, what can you do?

"No problems for a spare room upstairs tonight. If you're looking for some cash or any work, the Greek gentleman has the best real busy hotel-restaurant just across the road with staff rooms to stay in. Heard he offers a pretty good wage!"

"Get out, Are you serious?" I ask.

"Put your army training to good use as you did at the Yellowknife INN of the Dene Nation. Now this community is smaller but sure friendly. As it's on the main highway, the right hand to Foisy Ville AB, so always a busy place with two nearby Aboriginals reserves, one of the Blackfoot/Metis lineage, to check out. Many have our family names also."

"My mother is a Foisy I'll tell her about that it will make her laugh," I said. Well good golly as we cheered the cold beer he told me that the band playing that week was awesome. The chaps were very tall, said to be the tallest band in Canada. Thank God, what a day.

A quick walk next door to check on the car's status and the man told me about some welding needed as well as flushing with a new hose and a few toppers of oil. He said it would be as good as new in a couple of days. Okay right on, then told him where I was right next door. No sweat you know where we are. Pay up and the keys will be waiting for you.

Sounds good. My prayer was answered for a workplace just like that with a car repair at decent price. They had staff quarters that were empty, with a cot.

Went for a much-needed sleep at this hotel room, and a chance to think things over.

Later came back down and called to see what my Quebec buddy and his wife were up to this fine spring time of the year. Finally got a hold of them to see how they were doing, and told him I had no transportation at this time and I'd had to sleep with a beer at the pub, so see you later. I was near where the bus picks up people, next to the gas station, with rooms and a bar with a great band or so I am told. He asked me how long I had been in town. It was a big laugh when I told him, just this morning and we would catch up later.

"Come on over, we have some moose to eat and cut. Hahaha."

Well sure enough got a ride over to their house where he had a chunk of moose for a family feed, and we were soon catching up on the latest. He and I had fished, and hunted in Yellowknife for caribou and moose. Upon my arrival he stated the freezer was getting a little low but that a family member had caught a moose in the bush not far away so we could pick up another chunk to clean.

Sounds good to me let's do it.

Well, two days later, I went to pick up my car and I pay the bill and mosey on over to that Hwy Hotel restaurant run by a middle-aged Greek gentleman who told me that my friend across the road said I would be stopping over. Small town, eh?

Showed him my resume with my military certificate as we sat down. He had a key for the room that came with the position with a piece of paper on wages overall. Once again we shook hands as he told me that this was a busy trucker joint which was open from 6:00 am till midnight with a sort of a highway menu. If we have it never say no to those who ask for it – eggs at midnight or pork chops in the morning. How does this sound to you?

No worries at all I look forward to all the hours. Sure glad my friend told me about this place, it will cost me a high five game of crib and a box of beer. "

"I heard you were looking for many work hours to later take your meat-cutters course in Ontario this fall," he said. "Good golly, take as long to a lay-off as you want, after you show me I can trust you. I need a break myself. Heading down to Edmonton for a visit to family then all the way in Las Vegas for a month or two."

No problem at all telling him I can be trusted seven days a week till he gets back. His nephew and niece were in town should there be any difficulties. Bills and paychecks were all taken care of just needed the order placed in so you do not run out of food, using your military skill.

"Alright, sounds good. Will do the best I can not to let you down."

"Good luck in Vegas."

The work was on a smaller scale actually than feeding large crews. I was not alone; there was plenty of experienced staff to all make it happen. What an experience this was to be, trusted by this Greek gentleman to honour his reputation in feeding paying customers in the quality style he had employed for many years. Right on.

Read the menu to memorize the items; mostly names of certain items overall. It was just a piece of cake. The next morning he introduced me to most of the staff members, along with his contact relatives. I got a key to the office with an open tour of inventory, and the important points of the all-around business.

I put in my time on split shift mode to do the best I could, and was near the phone at all times. With access to the laundry room our clothing was impeccably cared for at all times. This was an awesome opportunity. Every day was good. Sometimes meal hours brought in a full house of busy times, I sure liked it a lot.

Here we go, sixteen weeks later he returns with a tan and he waves to me to come to the table for a break. We had a cold beer with a few of the staff later that day, and he told me to take my time in getting ready for that course I seemed to look forward to. Make sure to stop on your way back, you never know, okay. Well by golly my prayers were answered. Weeks after coming in with a layoff after his much needed break, here I was soon to be on the road again. Spent a few days at the awesome Jasper Heritage Folk Festival on August 3, 1985. On my way to Edmonton AB, I offered a ride south to a German hitchhiker heading out himself. I had seen him while he had been camping at the local park, and a few times at the local bar in High Level. Small world – he told me how he had gotten to High Level AB himself. He had hitchhiked out of Edmonton after having a cold beer with a local band he had made buddies with at a pub, then jammed with as they brought him up to play with them. Now that their two weeks' time was done he was on his way to this musicians' gathering where he was to reunite with band members from various parts of the world he had met some ten years before. Sounded good to me for a pit stop on my way to Edmonton. He asked me if I played any musical instrument. I laughed, stating yes, two years playing a trumpet at high school…my mother would send me practicing a mile away. Off we go to Jasper where on our arrival I got treated like a musician in thanks, with the opportunity to sing and hit the tambourine with them on the streets where we got filmed that weekend.

Jasper Heritage Folk Festival was a three-day weekend event and some musicians and crew were there jamming away the day before at a local football field living/using a tent, and enjoying the sunny weather and meeting one another. What a good break, listening to many bands with a place to stay with the many worldly musicians I met all introduced to me by the German fellow that had caught a ride there.

This gave me a place to stay for my three days there, along with singing and playing harmonica and tambourine.

Said thanks then goodbye to that musical group's organization.

Cruised non-stop and slept in my car for the three-day travel back home to finally see my parents.

Well here we go. Arrived in North Bay for my visit and had brought along my EI validation as an option to take that great introductory, meat cutter course. I went to the Employment office to show my work weeks, in hopes that I had enough weeks for this meat cutter's course, which would start next month, in the fall 1985.

Wouldn't you know it, I was still short four weeks along with the fact that not all jobs qualify for provincial opportunities according to government standards. However, luck of the draw, someone at the Employment center front desk said with a big smile, "We have a four- week tobacco picking availability. As you have past experience twice before, does that work for you?"

This was the end of the season so it validated EI's requirement for my twenty week course. "It is just a short bus ride to Tillsonburg. That's right, you already know your way there."

A Dutch family was waiting for my arrival to take me to their site with all accommodations, for a short four weeks, to finally get my meat cutters course. This family had sheep and I was shown how to shear them and to also de-testiculize the male sheep that were chosen.

After ten days we got to have an amazing stir fry with a bottle of wine and cheers for that event. They treated me very well, like being one of the family. I picked many baskets of tobacco, yes enough to make me quit smoking, after telling everyone that it sure was worth quitting now going on eighteen years. I'd shaken hands with Stompin' Tom Connors back in 1974 at the Tilsonburg Hotel, and enjoyed it all.

First language being French was part of the agreement to take this meat cutter's course. Upon my return, as part of my application I required a full medical head to toe, with blood work and a stool sample to check for any infectious diseases so as not to contaminate all meats handled. I had to give them my resume as well answer questions about any related background in that field. My military cooking skill along with my experience with hunting in northern Canada applied. Being born and raised on a farm in Ontario, I had learned a little of it and it sure paid off big time, as my dear loving grandfather had shared his skill at every fall harvest with all our family, and we'd all had a small role in it. Wild game often gave us a nice new taste for our supper.

Well I successfully completed the six-month course and had a few job offers but not at the wage expected. Truly as a graduate, I needed more hands-on experience. Although I did get offers out of town at slaughter houses in southern Ontario. Should you want to move there you would be hired right on the spot? Sure was true.

CHAPTER 14

FRIDAY, THE 13TH

MÂS'KÉG MIKE put in many safety hours for the next ten years in Ontario, Yellowknife NT, Nunavut, Yukon and British Columbia I cannot tell you all my stories here a few short one at the many sites and communities worked at visited in between also drove taxi in three of them.

DOMCO FOODS – 1986 (A)

I graduated with thirty others from an amazing six-month, meat cutter's course in 1986, and then I got to cruise around looking for work with a few other meat cutter. Our class donated blood to show that we were healthy enough to be hired for the pride of our certificate. We put in some time and our resumes at few chain grocery stores, and slaughter houses. Applications were put in at all places in the surrounding area of North Bay. We were looking for a good paying, full-time position with the skills we'd learned. Got visit my uncles in the big city of Toronto, then the Kitchener/London area. Yes these positions were in big demand, should you be able to move away from home to a good-paying job. But rather than move away I wanted to stay in North Bay to be close to family.

Good luck to my friend's career adventures, as it looked like one had come my way. When I called home to touch base, my mother said, "Get home now. A fellow called on the phone with a job offer and he needs you right away. Dominion Catering wants you."

Big Jim, an ex-football player, was our boss at the North Bay main office as we discussed the contract offer to feed a group of 150 men and women for a twelve-week period with a crew of two ladies and two men. This site being a future mining area, it was open twenty-four hours to feed crews on three shifts, with take-out lunches. $110.00 per day with a bonus upon completion of contract with inventory list. If all went well, food cost was at a proper level, and the crew gave good comments I would likely be sent back to the same site.

Although the camps were bigger than the ones in Yellowknife, by gosh sounded good to me. Handshake and then was told that I was hired right off the bat, and to pack a bag and grab a flight out of North Bay, going to Thunder Bay, Ontario on Friday June 13th 1986. Then the following day I'd be flown from there to Pickerel Lake.

Departure time on that horrible Friday the 13th it was pouring cats and dogs, with lightning shooting out of the sky. Sure enough a ticket was waiting for me on that day as I arrived with packed bags in hand and full gear all set for a twelve week stay. I'd just had coffee at Tim Horton's with Big Greg to sign my contract agreement. He'd said hello to my parents and off he went from the airport, as he told he would come out for a visit near the end. Military skills sure paid off, eh! He told me I would be running the show while feeding a large mining crew of 150 men and women, way out in the middle of nowhere, sleeping with access to a rifle and bear mace already at the site according to the standards of the group out there.

With my parents, sister, and nephew to see me off, we waited for acknowledgement that the weather would allow us to fly. As they were all crying, finally the captain give us the thumbs up, and said yes but the funny thing is that you're the only one on this flight.

The sky was sunny blue once we got past the clouds but there had been cancellations by few others. The crew smiled and said there would be more breakfast plates for us all. As I entered the plane I saw that yes I actually was the only passenger with this smiling flight crew. We flew out to Thunder Bay without any difficulty whatsoever.

Overnight in Thunder Bay, stayed at the marvellous Valhalla Hotel, and ate the biggest pizza ever of my life, which I shared with two others before the morning flight. On the way to Pickerel Lake that morning, I still had a few spare pieces of pizza, walking around to hook up at the Airport for transportation we were told that Pickerel Lake was located out on the Trans- Canada. If you were driving a vehicle, it was about 300 miles north with a right-turn at the community of good old Ignace, Ontario. As we walked by the entrance-way there was a couple with family having an 1600s style wedding, with a horse and buggy outside. At the beginning we thought they might be filming a movie what with all the cameras, but we were told they would let us know as we went to the outer section to sit and watch before the airport vehicle arrived at 10:00 am. Yes we were part of the crowd and likely made it in a movie clip after all.

It was just amazing as we watched the wedding end with her long dress being carried by a girl and a boy before the couple, in British-style dress made off to the outdoor horse and buggy with two uniformed men making the sound of a conch seashell using a long horn as they departed. Sure felt like being in a movie.

Then not long afterward, our airport limousine arrived for our departure to Ontario's northern region. We made it to the airport to fly out to Pickerel Lake, where we were then picked up and taken to a hotel for the overnight stay by a Mr. Steinko, who told us the departure time of the smaller plane at the dock area. He was our in-town contact should there be any personal needs, emergencies, or what have yous. This being one of those days, he smirked at me some in a private discussion area telling me about some issue that had happened at this camp. We would be flying in with an OPP officer to make an arrest and sort out what had occurred. There would be assault charge in regards to serious scrap that had happened to one of the staff I was replacing.

Initially the boss told me to listen, pay attention, and make notes for a radio call later that day. He said that we needed to take care of incidents like this so they don't happen to others in the future. Then he told me to take it easy.

That morning we flew in to "Hour Lake" aka Pickerel Lake and the OPP officer asked me to talk to our co-worker as part of the 5W's to help both our reports. I was to confer with my boss by radio as he was later flying in himself to assure all services were in place to feed the mining and exploration crew of over 150. Within the hour, we landed at the small, twenty-foot wooden dock to start unloading my gear with some food items. Two fellows stood next to one another with a bit of a grin not far away, as the officer and I nodded to one another. The fellow on the left side was my co-worker there to tell me what had happened. As we shook hands in a more private area, the story unfolded about the events that occurred, which were alcohol related.

Apparently, this Quebecker Frenchman's toupee had mysteriously disappeared during the occurrence of this life-threatening scrap. The cook I replaced, had some sort of a disagreement with one of the workers of another contract crew; something to do with the food prep and the kind of magazines he read, along with drinking alcohol in the kitchen, which he was not allowed.

Shortly after, while on the dock site still, the local OPP found his man, arrested him, and put him in cuffs. We briefly spoke to him and then to the cook. He had a knife mark on his throat, which he had received in the scuffle. I shook hands with OPP and I asked him to kindly send me a copy of this haywire story. with more details to be passed on when my boss flew in later today.

The gentleman I was replacing nicely stated someone had stolen his toupee but that he had no idea who. Speaking with a bit of a fat-lip lisp injury, he said, "Good luck buddy, they're a rowdy ruff bunch to feed," as he waited to hop on the plane to later press charges at the OPP's in-town office for the injury sustained the day before.

I helped them load on the plane for the flight back. The OPP with the assailant hopped in the rear seats and re-cuffed his man, as my co-worker dragged himself up front. I grabbed my gear to head to the Diamond Drillers office just up the hill, about fifty feet away from the dock.

As I made my way up the trail to the foreman's office, some of the workers whispered, that I must be the new cook and expressed a hope that I was better than the last one. I waved at the head of the drill team, who said that a few of the drillers knew me from Yellowknife and had said that I was a damn good cook, so not to mess with me. After an introductory discussion with the foreman, he showed me my quarters.

"That is the canvas tent next door to the kitchen and here are the freezers to keep an eye on. I see you have your twelve-gauge shotgun in case there are bears. CAF trained, for safety on location, he had a cabinet for me.

I dropped off my bags at my sleeping quarters then headed to the three-tent long kitchen where there were a good twenty fellows of various trades; some geologists and line cutters from North Bay. All of a sudden, a toupee comes flying at me with a loud cry of, "I hope you're better than the last guy, eh!"

As I held the toupee in hand, I turned around and nicely said to the young fellow that my army-trained chef skill was the reason I was there. I added that I had helped one of the driller's helpers from this current site and that I could lift a hundred-pound drill rod with one hand, so he should go bring the toupee to the gentleman that owned it or he would be on the next plane himself. Off he went. As we shook hands he apologized and I told him the OPP might ask him a few questions also.

By the time I got settled in the cook's quarters, I had met the four other kitchen personnel. On the night shift was Rosie, a very nice Aboriginal lady originally from Saskatchewan. She told me she liked it here a lot. Then the bull cook, aka Oh Henri, a stout Manitoba man with big smile. He cleaned the three outhouses, fuelled us up with diesel, and emptied the garbage from the next plane into its bear-protected area. Then there were two dishwashers. One was Elvira, a middle-aged lady that could wield a paring knife better than anyone else. Her nephew was a helper to keep the place clean and in great order. There was a menu in place for the hard-working crews. I then received the grand tour of the 150-man camp; TV room, offices, and mechanic shop. Turns out I knew Rick the mechanic from other bush sites. Sure was a small world.

It was getting to be supper time, and though it was June the weather was nice enough for a BBQ. I checked the freezers; three large ones set up by the kitchen tent. Well what a surprise. All three had been unplugged for a week by the looks of things. There was a lot of thawed meat on the top, thankfully the bottom was still frozen, so we checked and plugged them all back up. Checking the meat properly, we found we had a few good meat dishes before any spoilage. They cheered me on I asked them to call me MÂS'KÉG MIKE!!! I was sure ready to put my twelve weeks at this site without fear.

Like many who ever got nicknamed for a reason, I didn't mind this one, as there were quite a few other Michael's at our site listed on our schedule so this worked fine.

I pulled out all the thawed stuff. There was a lot of beef but with a crowd of 150 people, steak is usually popular. So we cooked all that was thawed and had some lunch meat and more left for stews and soup; things that could be refrozen for use later on. Over all it was not too bad.

"Sure hope you're better than the last guy."

..........My first stretch was from June 13 till September 1986.

The role of a cook is different in isolation than it is in town. Sometimes you're a morale booster who keeps things cheerful playing cards and displaying a good attitude. Or you're helping the team on a break, setting up camp to make your environment more comfortable and taking personal orders, communicating on the radio, or going to the movies.

Generally you're living and working in canvas tents, depending on the time frame, or the size of the isolated camp you're at with the crews. Important to remember is that professionalism and respect are key in this business. Long periods of isolation and many hours of work can make people irritable, lonesome for home, or stressed about personal issues. Being a good listener is helpful. Don't give advice – it is better to remain neutral. Should there be any injuries or conflicts, report it to your supervisor ASAP.

Many folks all over the world have a reputation good or bad, and along with that maybe a name like mine is given:

MÂS'KÉG MIKE or Frenchy.

A few of the great ones I have replaced or worked with are:

Rosie Pose - Twisted Sisters of Saskatchewan - Pirogi Paul - Macaroni Marcel, Art D Head, Down Town Dominic – Guy Have an eye, The mad Greek and many others.

Drillers called: Marvellous Marvin - Digger Dag - Hector the Director, Knivan Ivan- Laval it All, these names were given to these hard workers with as they did very well regularly.

BATTLE OF THE SEXES (B)

I sure enjoyed the first twelve-week stretch at a future mining site near Pickerel Lake, Ontario, along with a decent paycheck and a little bonus to say I covered all the golden rules. Now on my second stretch I was arriving again with more confidence or so I thought. Many of the crew I had been feeding was now part of a team that was elsewhere, and there was a whole new group of staff. After my much-needed two-week break, I was looking forward to going to the same location for another twelve-week stretch. I was in charge of a whole new group of staff, two women and two men and this should be a lot easier now.

Here now after my two weeks off, this future mine site was officially at the next step that had something familiar to me this was going to be the first blast at our site. The where setting up a second entrance way as I had seen done before at another mine site. The first blast to the shaft on the south side entering of the future mine site. Holy Smokes! Rocks flying all over the place with many hiding under the trees far enough from our quarters. What a loud noise with a scary show of flying rocks perhaps coming your way. Yes before the blast, there was word of mouth, warning horns, and double checking by radio. Safety first. It was all over in less than a minute with smoke coming from the ground. No one was hit by a rock, phew!

This was a reminder that working in the mines even as of today still can be dangerous on the surface, and even more so underground. Time will tell how good the find of this mine was. Where the industry has established itself in various places all over the world, Canada has become very well liked. All is said and done, now off to the kitchen for the feeding of a large group with a list of each company and their employees. The head of that mining company was there with a staff including a geologist/prospector with six apprentices, diamond drillers, a twelve-man crew, and many tradesman, as they were starting to build the actual structures. The work time in Ontario is also influenced by forest fires that allow you to keep just a small group at the site with local Ministry of Natural Resources (MNR) stopping over. Pilots fly in to camp with new supplies, every second day, or with new orders for crew changes, or to bring patients to the in-town hospital if need be. Often a helicopter is at the site for an immediate evacuation. All of a sudden we may all be asked to leave ASAP due to life threats in the area. Yes indeed this is where Aviation - MNR/ Forest Fire Rangers(AFFES) are found, sharing knowledge of the fires and taking care of them quick as can be.

Here now going on my sixth week in, with all aspects covered, staff doing very well, and no complaints, or so I thought, suddenly there was a rumour that someone was abandoning this crew in a short bit of time. Phew, what excitement there can be, even if away from the big downtown or where ever you are from, eh? Here we go, as we continued with daily activities or so I thought. Then, the two new ladies at our site approached me with a petition from of the on-site workers. It had been signed by a few of the foremen and had a total of fifty signatures. As we sat in the kitchen with a beverage, they just straight up told me, MÂS'KÉG MIKE, that they were now the bosses and that I was being fired today. Yes, pack up your gear and get ready to leave on the next plane. They told me this was all their idea and that they wanted to take over to run the show, so that when the mine opened they had the forms to prove a contract offer. So having been at the site doing a great job, I just confirmed with those in charge that this was all true, and asked if the ladies were having sex with the head leaders. There were a few smiling faces to answer my questions. I just packed my bags and flew out to Pickerel Lake on the next plane out, then called my supervisor, Big Jim Craig, as we did not have private access to phones out on them there isolated hills where the mine might soon be opened.

Yes indeed, within the hour I get called back by our Dominion Catering boss with a wee bit of a laugh. No worries, just have a beer and stay near that room till I get there in a couple of days with a replacement crew to replace those two ladies. No worries, run a tab and watch some movies. We have an in-town storage unit – kindly check it out. Well with all said and done, I did not mind what the ladies had pulled off for whatever reason, because it was a nice little break for me.

Sure enough, a couple days later Big Craig, our ex-football player who is and impressive 6'7" and 300 lbs, arrived in good humour and upon his arrival gave me a firm hand shake. Got the other two fellows a room at the hotel telling me that one of them was going to be night shift and the other was good on dishes and had all-round experience. We were all flying back to the site tomorrow with the next load. He asked me to tell me a little more about what had happened for tomorrow's approach. Isolated workers are hard to keep steady so those two ladies will get a short break of course, and a vocal disciplinary reminder at our discussion that I am their boss and that their paycheck comes from Domco Foods. Notes for a report in and we had cold beer meeting before early departure on the morning flight. This was a reminder of who is making your paycheck, and who can kick your ass out of this site.

Yes indeed, another sunny fall day with an enjoyable flight to "Hour Lake" site with a load of supplies now often brought in every second day. Where I am told that a group of shareholders will be attending a few days in the near future, so to keep things A-One for the bonus pay. Food costs will improve with items from other nearby sites, which will be sent to this one, so inventory done once a week will work for you. So here's the deal you go up to the kitchen, call both those ladies in for a staff meeting and do not tell about me just yet, as I'll be talking to the others who want to be boss of this site...the ones who signed. Everyone has the responsibility to work together with all the head bosses coming in this week so cover all your bases with good choices of main courses, soups, desserts, and luncheon stuff. I said yes boss, shook his hand, spoke with the two new fellows along the way telling them what the site was about and let them know it had the best fishing of pickerel ever. They liked it all. Asked them to watch how this is supposed to work.

So now this being my turn, I nicely walked over to a table and I grabbed a coffee with my page of notes, with their names on it only. They asked me what the hell I was doing back at this site. Stated that I was there with good reason and offered to sit down with them for a fair discussion to let them know exactly why. Just asked a few visitors to give us some private time together for a meeting.

Of course I did it short and sweet by stating that a few new fellows sitting at the far end were here to replace them, and that they were not fired, however, would be transferred today to another site, or so I was told by our boss. When they walked up to me to ask about who was this boss, all of a sudden the big guy walks in, as he was listening at the door. Oh by the way our boss did fly in with us as he spoke to folks in charge here to let them know what is now taking place. You two are getting relocated, or you can pay for your way back home. That is the way Domco Foods does things for all employees of a decent wage. I was given the opportunity to let you ladies know to pack your bags now as the plane is waiting. You are flying out today within the hour. Yes, I did finish off my twelve weeks for about a $7,000 paycheck with a bonus to get ready for another site that was just opening up in the New Year.

Screaming Eagle - (C) 1987

New Year of 1987 I was flown in to the mining community of Red Lake, ON and told by our pilot that our last site of Pickerel Lake, Ontario was just fifty miles away. This being another isolated area with high potential for a future mine site that could just as easily be there in years. At least they know that there is gold in them hills for the stock market manipulation. Big Craig Jim had to use a land line through a separate geologist camp group, as co-worker Olga had items for her eight-man camp to pick up in town that day, MÂS'KÉG volunteered to go with her for the a one-day flight in to help pick up personal items and groceries for both our work crews. Here we were back in town with everyone too busy with a task for each of our own team –the weekend rush hour of a small town even in Red Lake, Ontario. I took a taxi and did a pickup of all our commissary items, including a trip to the liquor store for a beer supply as we were on our last week for that site, or so we thought. For sure a new six-week crew was heading out.

Olga had been living in the area for a few years, and knew a lot of people only too well. She and I toured around town for the shopping list, double-checking for all items. Of course we returned to the helicopter pad to load up the goods ASAP. We were staying behind till early the next day to take a break after five weeks and then would fly in with the plane with supplies on the next day. This was a chance to visit and shop around at the grocery store for items we hadn't yet found. What a pleasant surprise, now here we are. Because of a brutal overnight snow storm the next day, we got stuck in town for another paid day. Oh well. Just too bad about the weather. Olga and I had to share a room (purely platonic) for the use of a shower, and to do some laundry. We did go out have a few beers to cut loose at a local pub called the Screaming Eagle. I got to meet some of her good friends – she had been living here for a few years now. As we enjoyed the band and spent the night dancing and acting silly, Olga announced that she was sleeping at a good friend's house so I had the room all to myself just in case I might get lucky. Humour was always part of isolation as often your workday could be your last one.

I closed the bar and called it a night to get a good sleep and catch the plane the next day. What a surprise – it had snowed a little but the expected forecast was now going to be twice as much of a snow fall for the week, making it near impossible for a fly back to camp. Everything seemed to be put on hold. What can we do? The young fellow that was helping me said it might be just too much for him all by himself, so we gave them a call to let them all know. Forty hard workers to feed around the clock.

I spoke to Fast Freddy, our head camp man, to tell him the bad news. "Ya," he said. It was snowing there too but he had an idea to suggest an alternative route because there was no way Super Dave the bull cook could feed the large crew alone for another day let alone a week. I agreed to his offer, which was for me to take a taxi south down to Ear Falls, then make a quick left once there, on an old road all way to the end at the South Bay mine site. Carlos, our camp guide/trail blazer extraordinaire would meet me at South Bay by Ski-doo. I was to make a call to the radio contact out of Ear Falls to confirm my arrival. Roger that. Okay, we were on it, I said.

The cab took me to the abandoned South Bay mine site not a moment too soon. He was afraid of getting stuck on that ten-mile road. I paid him and we shook hands as I trustingly sat to wait for over a few hours. Then, sure enough, there was a light coming right at me. He being ex-American military, it was no worries at all.

Thank you, Lord. Yep it is Carlos with a solid wooden sleigh to take me back on a two and a half hour, fifty-mile ride to the camp site, on a dark, stormy, snowy night. "Don't worry," he said. "I know these trails like the back of my hand. Here now, this is a flotation Ski-doo suit to wear, just in case we hit open waters okay?"

I laughed and gave him a high five, and that is when he pulled out this plastic bottle from one of his coverall pockets. "Want a snort of this?" he asked.

"What is it?" I asked him.

"It's home brew. Matilda handmade it with a smile and it sure packs a wallop."

Well alright why not.

"Okay," he said as he laughed. "You can have this one, coz I have another one." It was not mixed and tasted like, phew, the Everclear I'd had sample of in Alaska a few years back.

We were moving like the wind at a good clip and a little under a couple of hours later we pulled into where Carlos' secret island, year-round homestead is tucked away. "You get the full tour," he said. "Not many people do get to, so kindly don't tell anyone I live out this way. We're less than an hour from our site. I will show you why and where I brew the Matilda, with a take home bottle of my home-brew mistress for yourself."

Wow the place looked like a second-hand backyard shop of stuff all over in, this old, decrepit, hand-built shed made out of ragged barn board. There was a homemade still, four feet high with a ten foot width and a capacity of about five hundred gallons. It was all really no big deal as he told me the composition of it all. With a couple of cases of empty bottles and an outdoor five- gallon steam kettle with a gauge on the top, ten yards of copper tubing were coiled then lightly angled to sit off the propane element, which led to a waiting, empty, forty-ounce bottle. After the full display of the setup, Carlos put his finger on his lips. "Shh. No one knows or needs to know. Come, I'll get you one."

There was a brand new batch ready to just gurgle away in an old style 500-gallon, wooden whiskey barrel with the steel rings, and a wooden lid. A canoe paddle ore cut short was lying there, ready to stir the mix now and again.

"On my next day off I will make it happen. What do you think, MÂS'KÉG Mike? Here is a bottle for yourself. Good for the cakes," he said as we headed back to the Ski-doo. We were only ten miles up the way.

Yes sir, his six weeks were up, and Victor the Dictator was heading back home to Toronto. There would be another driller flown in for his stretch tomorrow, and other people coming in to make the flight work out cost wise. The bull cook and I had been conspiring about this day and while Victor was distracted and saying good bye to the rest of the gang, Super Dave stuck an orange latrine/shit bag in his gear as a going away package pay back. When they were departing in the helicopter with his buddies, I told one of the helpers to let him know I packed Victor a hot lunch for the trip out. He even said thanks. Wasn't till spring that I caught up with one of the guys at another camp. He was telling me they made it to the airport and headed in to wait for the connecting flight. After a while sitting down in the terminal the Dictator was looking under his cowboy boots checking for something smelly, and one of the helpers piped up and said, "Oh yeah, the cook said he packed you a hot lunch." He knew right away and unzipped his bag. Fortunately for him the bag had a knot in it and nothing leaked out, just a strong odour of fish guts with a sticky note that read: Bon appetite, to you, pot licked individual.

FORTY BELOW (D)- 1987

A crisp morning to be sure, at Pants Island with seven weeks behind me and ready to do as many as needed. Back to back stretch just a short flight away, as this was still part of a two claim package side by each staking allows one person to claim only so much at a time. We are out looking for a copper mineral deposit. Living in canvas tents can be a little brisk during the winter months, when you're getting up first thing in the morning. Even if you have a five-star sleeping bag, with snow banked on the sides of the tent, and a diesel furnace working at peak efficiency you can still see your breath. Yeah she really puts out and if you're real lucky you may get a winter liner out of the deal. The one-inch plywood flooring might not stick to your feet when you jump out of bed quick enough first thing in the morning. A few workers kept a pee bag or pail inside by the bed, good enough until one got dressed for the outdoor needs.

The ice road track was designed to transport equipment in preparation for later hauling the drilling equipment to the geologists' winter location for a potential next mine site. My coworker, poor Mr. Carlos, had himself a heck of a time. A week later as he and I showed up at our main camp, he was covered in ice and frozen in the sitting position behind me on the Ski-doo. I had driven him back to the main camp for the immediate first aid he needed.

The yelling sure got my attention, as we pulled up to the camp. "What is it?" I asked. Carlos looked out from behind his Halley hansoms floating survival suit, with a frozen smile and said, "Yaw I'm still alive, just get me out of this suit so I can talk to Matilda." We picked him up in the sitting stance and carried him to the shower for a rousing moment of warm water till he could move enough that we could get him out of the coverall. Then into the hot shower for ten minutes. Carlos laughed and said, "Gee whiz, can't remember when I last took a shower, usually I wait for the spring thaw."

A true bush man knows how to conserve water in the winter time, I joked. Yes sir, we bundled him up in a blanket, and set his clothing to dry in his change tent. Then we put him buck-naked into a fire blanket with his moccasins, so the sixty-five-year-old Carlos could make his way to his tent. Where we tossed him into his five-star sleeping bag, with a towel around his head for protection against pneumonia, he smiled and asked us to reach under the bed for his bottle of Matilda but said to watch out for his army pistol because it had a sensitive trigger that would wake everyone up from night shift. Shortly thereafter, I picked up the first aid kit for cuts on the forehead that were not deep, so I cleaned them and taped him up for a nap. He had slipped and fallen off the piece of equipment he drove to create a path on the lake for the next hole in the ground.

Then I went to see our boss to let him know Carlos was taped up good and the injury was not life threatening. That's good because he did not want to go to town as he apparently had a warrant out for his arrest for something. After a little rest with his honey he said if I'm not in for supper we'll see you at breakfast. I'll stop in later to keep an eye on him.

That is all I have for your report but that first aider rescue ride from the site sure put excitement in the day.

A couple days later Carlos is back on his feet, keeping busy around the yard, running equipment and tinkering about as I changed his bandages with a clean wipe. He was wearing a big grin under that gray beard and handlebar moustache of his. "Hey," he said. "Heard the workmen on the drills like moose. I got ya a big roast. Do you want me to bring it in?" he always asked with a smile.

"Okay," I replied. "Bring it in."

All of a sudden in he comes with a whole side of a moose calf, and tosses it on the floor. "How many steaks can you get out of this?" He laughed. "Compliments of the Cat Lake Reserve Chief's daughter," he said. I cleaned it up good and made some roasts. Word of mouth of appreciation for wild game was seventy-five percent. Okay for that, and pork chops for the rest who did not want any, or macaroni casserole.

Randy, a good driller mechanic with a reputation as a very safe worker from Manitoba, came calmly into the kitchen this mid-morning. Carlos and I were having game of crib with a coffee, enjoying the day. Randy very calmly asked me where I keep the fire extinguishers. Well, one by the door, and another right here by my propane stove area, I pointed out. "What's up? Do you need the use of one?"

"Well yeah," he stated. "I found a chunk of ice in the diesel tank of the go tracks Carlos runs so I have a propane torch hook up in three feet of stove pipe to do some melting of ice, and it just caught fire."

"Out there right now?" we asked.

"Oh yes," he said.

Well Carlos and I grab the fire extinguishers and buckets of water and through on jacket and gear because it was forty below, to have a look see. Good golly Miss Molly that is sure a fire! We doused it all around with the extinguisher and threw the water and snow all over as the diesel was on fire.

As we were heading to the end of my fourteen-work weeks, with spring on the way, the ladies heard from a few that black bears had eaten their cat, so they were moving their tent a little closer to ours. They asked if we could also feed their crew of ten as the head geologist's wife Elvira was heading out and she was scared.

Carlos Vinous, was a camp trail blazer with a great deal of respect for local knowledge, a part-time guide, and friend of the local Cat lake Reserve, was known for the home brew of the infamous Marvellous Matilda.

Having the skills is not enough; I needed the ability to last the duration and come out with as little food as possible, ordering enough for the week ahead during operations, and trying to please the fussy people who liked name brands and had their preferences of culinary delights. Keeping it clean, simple, fresh and hot, checking for allergies, likes, and dislikes.

Because you may end up with a bad name, hah hah.

SHORTY, THE BULL COOK (E) – 1987-88

I was friends with a few named shorty, here now I had got to meet another Shorty in Pickerel Lake ON. There are a few people who have this nickname of Shorty, in particular, and they have great reputations combined with a good background that is well liked in society, and in the workplace. I have met more than one with that infamous well-liked description.

Shorty Sundholm and I first met in Pickerel Lake and we became good friends. A kind soul with not a bad bone in his body. He claimed to be a stubble jumper from Saskatchewan. I guess he would have been in his mid-fifties at the time, a kind gentleman with a great sense of riffraff humour. His favourite saying was, "I'll never tell a lie unless the truth doesn't sound good." Then he'd wait to see what would be your reply and then start laughing when he knew you caught on. There you go, as told by a Flanders for many generations.

He was the bull cook in our camp, an all-around guy, and he could sure play a mean game of cribbage. If something needed to be done indoors or out, they would say, go ask Shorty.

Bull cook: One who does caretaker chores and helps the cook, or in small camps is the only cook.

Mâs'kég Mike had a cribbage reputation to uphold; my average was eight out of ten games in competition, making me one of the best crib players in camp yes, that's in every camp. If we had to do something in between games, we just marked it all on a game pad.

Rivalry amongst co-workers, drillers, geologists, and pilots, was sometimes fierce. We would form teams to compete against each other when time allowed. Usually it was in the evening over coffee and some pastries. I only ever got the 29-point hand once in all my life. Shorty had three 29s under his belt. He and I used to team up against the drillers or anyone that would dare to take us on. We usually got to play whenever we were on breaks, after supper, or after the completion of the day's duties. Mâs'kég kept a list on the fridge of who owed what and would challenge anyone when it was not too busy. Usually we paid up at the end of the run or when a person was leaving. We respected the rules equally and paid up before the person left or on the next turn around as agreed. We played competitively, to have fun, and to have bragging rights for the day. Sometimes cheating might occur accidentally in the points or by not being aware of the actual count of points in the hand itself. According to the rules there's no helping your opponent, as others watched in fairness to the players. Yes rules for some allow you to steal by pointing out the missed counting of your opponent.

Crib is a card game played with possible combinations of 15. It's skilful and witty at the best of times, and with good bluffing ability and a little luck on the cut can be played in a variety of ways with as many as four players, preferably enjoyed with a small wager.

The bull cook duties were to hand pump diesel in the camp, for approximately thirty-five tents. These duties also included pumping water, shovelling snow, burning garbage, mucking out drillers' gear and dry storage, cleaning latrines, and occasionally the odd kitchen chore like dishes or peeling potatoes if the cook was too busy. The cook is usually up at 4:30 and the bull cook is in for breakfast around 6: 45 am because the seven to twelve hour shift starts at seven o'clock sharp every day. For safety in the workplace, this is a time for discussion with the foreman on the day's events; reporting problems, or medical/accidents issues at drill sites. There is troubleshooting of problems, breakdowns, and discussion of supplies needed. Everyone has their list ready to coordinate with the plane, which usually flies in once a week. When the plane comes into the camp they may call ahead for a weather report, lake conditions, or information on the thickness of the ice, which is checked out every day and shared with all at site.

Periodically there may be a radio in the kitchen as a safety measure, in case of a medical emergency and to save travel time. Nowadays there is a chopper on standby and hand held radios, as well as satellite-style phones. Working conditions have become safer, for everyone nowadays.

With Buffalo Joe all flights in the area are handled efficiently, and keeping in contact is number one for all.

One winter at another camp called Pants Island, we worked and played crib over a period of three months. This was a camp of about thirty people, and seven of them were young ladies training in the field. Though it was unusual at the time to have so many women in camp, it was a very memorable time for them. New to the field, they were sometime victims of practical jokes. That is part of the initiation for all newbies and it is also a great time for everyone. One day, two of the ladies were helping Shorty clean out the stove pipes on one of the diesel furnaces, to safely improve the heating and see how it is done. There was a pile of chimney soot a foot and half high near ready for completion, when all of a sudden the chopper returned to camp to get a load of supplies. It took the new drillers about fifteen seconds to be covered in soot. Well, Shorty hightailed it out of there, just a running. The girls did not catch on and stayed where they were, of course still near the soot as it flew up everywhere and they got completely covered except for the eyeballs. Oh there were some laughs afterwards and the pilot did not realize it till it was too late.

Shorty and I would team up against the drillers on the down days and we would play for cash or some bonus prizes. Drill companies used to offer nice prizes for safety records, like watches, rings, chainsaws, sleeping bags, and jackets. On this particular run the foreman had a big laugh. Shorty and I had won a dozen ball jackets, hats, two chainsaws, and one gold ring from Knivan Ivan.

All the guys rolled up their sleeping bags so critters wouldn't go and make themselves comfortable – kind of like the old rattler in the desert trick, only with mice.

Well after a bit of teasing, especially working in the camps, some people might hold a grudge or might be planning to get even. Most times it is deserved but sometimes people might get fired over it. Of course there is always that shrouded veil of mystery or the hush hush so everyone can keep working, save face, and not get one of the nasty names hah-aha, and we don't find out until we get to the next camp.

This is my only 29-point cribbage hand ever.

The day shift guys had played a joke on Shorty – they had put a mouse in his bed roll just before he went to bed, and he didn't find it funny. The foreman warned them, don't mess around with Shorty he will get even with you. On this particular morning, Shorty grabbed an old, leaky pot we never used and pretended to soak his feet while he was peeling the potatoes. All along there was a smaller pot hidden from sight. As the Drillers walked in the kitchen for breakfast, they looked at Shorty and smiled. "Hey Shorty, what are you up to old buddy?"

Shorty kept quiet. He had a serious look on his face and replied, "Have you never

Seen anyone peel potatoes and soak their bunions at the same time? Very therapeutic, you know." Potatoe water does have a lot of vitamin C in it, sure works for me, eh!

Needless to say that had the drillers going, and talk that Shorty might be upset got around real quick, as he pretended to peel the spuds in the same pot he was soaking his feet.

On another day, the helicopter flew in with the shift change crew with news that they had spotted a large grizzly bear from the air. With a big grin, they came over to Shorty in the kitchen and stated that we had better be careful, and carry bear bells and mace for safety. Thank you for telling us about the bear that's not so far away, gentlemen. So I suggested to my good friend Shorty that he grab the slop bucket and fill it up with water, then dump it under the fellows' tent. He made sure they saw him and they asked him what the hell he was doing… just nicely said.

"If the bear does come to our camp he will likely come here first as this is the second slop bucket under your guys' tent, so we have spare mace and bear bells for you as well. Oh and by the way he may come this evening so grab some tape to keep at least one eye open while you're in the bunk sleeping." Well here we go it is looking good

CARLOS AND MATHILDA (F)

The year was January 03, 1988 and there were about thirty-eight of us flying to a camp called Pants Island just a few hours outside of the small mining community of Red Lake, ON. Some drillers who had already spent a little time there in between other bush runs told me this community of Red Lake was a lot like a smaller version of Yellowknife, NT. The Gold Rush days sure bring liveliness as Yellowknife was only about 5,000 in population upon my arrival back in 1979. Anyhow just another area that diminished in population on its last days of mining as workers head to the next one to keep a steady pace in lifestyle, finance wise. Geologists search for minerals, by testing rock samples and then following with drilling projects. The geologist's final core sample results are flown out and analyzed for their potential mining or for at least more drilling. Here now, this was one of them as the industry was in big demand with a small group that did not mind being isolated for a good wage and inside info on the shareholders next thumbs up for public promotion. Being first in on that is great, however, it can be sad for some, though worth chatting about at the pub. With place to crash, I completed the food list at the chosen grocery store for all the items and a had separate list for personal items that workers forgot, but no alcohol allowed…or so we are told every time at the start of a new camp. So let's check it out the town, we so do get a chance for one night to go all around. Nearby were the Aboriginal Reserves of the Cree folks. Some of these folks are a pair of locals who give out-of-towners, a place to stay or you never know perhaps a spousal approached.

In its day, it had been a boomer mining town with at least triple its population, but there was very little going on now aside from some tourism with it being part of an Aboriginal area. There was a pub in town near the lake but all the excitement was said to be either at the Screaming Eagle bar or the Red Dog Inn which had a piano. The crew I had worked with the year before knew this and preferred staying at the "Dead dog" as they called it, hah-aha – more of a fun place to be.

Mostly the same great bunch of guys as last year from just after the Pickerel Lake run. There was our head geologist Big Jim and his loving wife Elmira. Drillers and helpers (eight total) for the two drill shacks of twenty-four hours. The head of the team was Brother Bruno and Fast Fingers Freddy, the head driller was Marvellous Marvin from Manitoba, day shift Newfy Go Go Gordy, a Red Lake gentleman, and as camp-man Bud de Spud with his German shepherd hound. Top man of course, yes indeed, was Carlos our camp scout/trail blazer. We were told listen to him always as the lake could be cracked or broken, we should always follow his flagged marked path if even just fishing.

At a separate nearby tent camp a half-dozen young ladies were taking geology training with a couple of new helpers sometimes called the "Greenhorns." Good listening along with the power of the pen to take notes still works well in isolation. Do not give the girls a hard time or you could be on the next plane out. Oh yeah, a well experienced Japanese gentleman who was our helicopter pilot and a few new faces. Myself *MÂS'KÉG MIKE,* re-embarking on another fun-filled adventure of feeding the crews, and the Midwest Drilling teams.

This project was going on for fourteen weeks we were told! They also had yet another nearby program with Olga cooking at the small mobile camp, doing geo-magnetic surveys to search for potentially new ground to be drilled in the future. That was seven peoples for a total of approximately forty some odd with a few that fly in and out for various reasons. Our departure time was decided the day before to ensure head count and of course seat count as sometimes there was a surprise to all, and someone else comes along.

We all took off at different intervals at the beginning of the day; some on the plane, some by chopper with supplies, and as needed on either end to coordinate the start-up of the project team's work with mutual cooperation to accommodate a smooth beginning to the schedule. I was on the plane and just before had taken my food list to the local distributor. I'd been advised of a two-tent set up, which would mean two table seating's of twelve at meal times except lunch, which was at noon to one pm and was normally soup and sandwiches in, out, or later at your convenience.

These people were experienced and very professional, but occasionally there was a bad apple in the barrel. This can go one way or another and must be handled with tact and flexibility.

Hector, the director, was a long-time driller with the company. He liked to tease people to see how upset they could and would get before they cracked up over his head games. This time around, I was the target of his verbal assault, so from the first day he came into camp, he would ask what kind of shit are we getting at every meal, and at every meal I would try to think of something clever to answer with. Super Dave, AKA de bull cook, the new guy who was helping me, could not fathom why this man was so evil, as he would always ask that every day. As a newbie everyone kidded around about what would make him tick; a psychological profile of humour if you will while in isolation.

Old Bud was the camp man, sixty years of age or so. He had come for many years, one of them as tourist hunting/fishing scout. But there is no hope in hell I'm peeling potatoes for twenty-five people he said. Bud took care of the garbage, pumping water, shovelling the walk ways or the trails. He liked to keep busy. Lady cook Olga was a nice local lady from about five miles away with a small crew. She would stop over for a shower periodically. She would pay us an unscheduled visit along with forgetting to put her sign outside the door of the only man's wash area. She was a voluptuous women with flaming red hair, almost as red as Bud's face after he walked in on her one day by accident and saw her fully naked. "It sure has been a while for me," he said with a smile. They had coffee together and laughed.

Super Dave was a nineteen-year-old local of Red Lake who had trouble emptying the latrines, all three of which were wood outhouses that had five-gallon garbage pails underneath lined with doubled layered orange plastic bags that had to be emptied every day after supper and then bundled up and tucked away in a holder for the next plane. So Bud, his uncle, I believe, who had done it the year before, gave Super Dave a set of coveralls, gloves, mask, and goggles to perform the despicable deed for the next three months or for as long as he could.

One day, someone put a pair old gloves hanging off the seat with the lid over them and waved at him yelling, that someone had fallen and needed a little help.

Everyone's routine is pretty much the same day in/day out seven days a week with a twelve-hour work load, except for the food choices, the weather, and the locations of the drill sites, which become a major topic – that's what it is all about. Distractions include playing cards, board games, and long discussions about solving world problems after twelve-hour early shifts. As a lecturer of sorts each one has a right to chat amongst each other with storytelling about other sites, workers' disputes and injuries, which are a large part of safety discussions.

Very important entitlement for some as if they are at 5 start hotel restaurant of choices.

With fourteen weeks in that area many things can happen, some good, others...well, bad.

FIREFIGHTERS OF ONTARIO (G) - 1988

Here now Domco had me sent on Highway 11 to the north head of Lake Superior just east of Lake Nipigon, at the community of Geraldton, Ontario to feed the firefighters at one of the main fire station of that region. Summer of 1988, in the month of May we are setting up the station for the training along with running crews all the way to the end of the Diefenbaker Hwy 584 right by the bridge where the road ends. Ministry of Natural Resources was throughout the whole are special in the summer when there are fires, they are very well liked in all aspects. We got to experience how efficiently all of these brave working folks take care of our land, help save lives in a very demanding job at various places in Canada. I got to see first handed on my six weeks in preparation with adjoining communities to prepare for flight service role map to one another. At the main station we had five Domco Foods workers feeding groups passing through as well, with a mere phone call to ask for twenty to fifty brown bag lunches that would be picked up by a driver and taken to a coordinate location every day. Once a week I would take a load of food, driving a van full of mail and personal items. Yes, I did come equipped with a charged radio in case of any mechanical issues, flat tires, or other emergencies, to stay in contact with MNR. A full can of bear mace, with a tooting horn, along with my twelve-gauge shot gun for a backup just in case my life would ever be endangered. Thank God this was a safe work summer for our crews.

Junior Forest Rangers (H)

Near Nakina, Ontario on the Wildgoose Lake just off the Macphail Drive area, MNR had an awesome setup for the Junior Forest Ranger. I had herd so much about the Rangers over the years here now my boss told me a five to six week stretch was coming up just a few miles away from Geralton. This being where our food access was available for a near by drill crew as well.

Providing transportation of the food to 3 location over a period of 6 weeks would get me a bonus pay if *MÂS'KÉG MIKE* was interested in such a challenge. My boss Big Greg from our North Bay office had a schedule in place of picking up food items at our Geralton distributor for deliveries to a Drill site just a few miles away near the community of Beardmore Ontario. Then a double load on the way back to Wildgoose Lake to drop off a second load for our Rangers 100 peoples to feed, to then take a break there with a checklist of items for inventory on the following pick up. Then before dark to head on out the Firefighters crew past Nakina Ontario just off high way 584 to the end of the road on the right hand side there would be a landing strip, helicopters canvas tents with a good size quarters for food service of that area. This was a sleep over so you could depart the following morning after breakfast to head back to the Junior Forest Rangers site.

Sounds good I will get a map some direction from the Ministry of Natural Resources crew at our home base site to give that food run of a day a good try, sounds doable so why not. Well touching base with co-workers by phone then the use of radio contact with MNR this was going to be a heck of a long ride, to make sure I had a few spare tires on these lengthy gravel roads. Stopped over one day for coffee as I ran into one of my Meat Cutter friends out in Nakina that was very surprised to see me when he asked what the hell are you doing way up here. One of our co-workers at Forest Ranger camp up the road was living here in Nakina I was picking up some personal items for him from his sweat heart on the way back. We shook hands had a few good laughs over coffee then off I went again doing my round trip once a week worked out hassle-free of course my fishing rod was with me at all times, as well preventative measure to the black bear of the region. So every week we fed these hungry

BACK IN THE GREAT WHITE NORTH (I)- 1989

Well now, a little break from the Ontario region with a busier time farther up north for a better pay, along with driving hack in between, was always a nice break of sorts. Shoot the breeze with a few friends along with a stop-over at the Strange Range bar for the easy listening of country music with a cold one. Not long after changing my license and health care card, I'd hook up with a phone and post office box, and that was pretty much it. Then dropping off a few resumes I was offered work in short bit of time after my arrival to Yellowknife, and told right off that my skills would be put to the test and that there was high demand for food preparation for a large group, along with a load of trustworthiness combined with team style ability in the isolated, furious regions of cold weather. Army trained I replied, as my certificate answered.

"Yes we heard from the Drillers *MÂS'KÉG MIKE* you are back in town. Time for a one stop shop drop by the office tomorrow."

Working in isolation sometimes people come in as couples. This is nice if you can get accommodations or you're in the bigger camps. Then it is just like being home, which is great, if you don't have kids. Especially in the summer it becomes a working vacation for some. Where the girlfriend or wife is the one who does the feeding of the group, it is all okay.

Sometimes out yonder with single ladies, yes, a little romance does occur and the girls get friendly with the right boys to both have enjoyment. However, sometimes it is to guarantee their position or rally for a better one. I learned this first-hand as how that works was all new to me. Thankfully, rules and regulations are in place for the bosses.

Some fine ladies might be looking for a permanent relationship. Why not? It is a natural part of life. Sadly *MÂS'KÉG MIKE* never had the pleasure, though not for lack of trying. As Big Greg, my boss, told me once – do no get too friendly or sleep with the staff, it could cost you your job or you may end up doing their jobs as well. Ha ha. I heard that can truly happen.

Sometimes the camps get bigger or bunched into one and the bosses choose the more capable cook overall, while oftentimes somebody gets the axe or is sent to another camp. This does not always go over well for some. Anyhow, here we go with the tale.

We were at the Sunset Camp just off the Beaulieu River system just seventy miles northwest of Yellowknife, less than a one-hour flight away. As in many successful businesses, in which the best bid gets the contract, our company was awarded the next one; likely for its reputation for that area. That particular sunny day my supervisor walked over with a big smile and he asked if I could do another six weeks.

Oh yeah no worries, right on. Little did I know the Twisted Shysters at the other camp were conspiring to get rid of me and take over my job. Because they got wind of it first. Oh well, let's wait and see.

As he returned from the Sunrise camp just two miles away, the foreman informed us of the great news. It was confirmed, yes, our outfit was awarded the six-month drilling contract. Then we'd be taken over to the other camp just a little farther up the lake with more sleeping quarters, to start within one week with a bigger crew. As all this happens very soon let's do the math and check for space and so on. Top of the list, I had to go over and do an inventory of food items from the other camp, check freezer space, and order for fifty people. We needed some extra items due to the ice break-up because in a couple of weeks there'd be no more landing on the ice for Buffalo Joe flights, we'd just go without. Keep lots of macaroni.

It was mid-April, a perfect afternoon, when I arrived there by Ski-doo, what a perfect day. We were soon to move in and take over, so just a roundabout tour with an inventory peek. Upon my approach, a lady with a forty-ounce bottle of rye greeted me at the trail entrance, then asked if I wanted a drink, calling me a bunch of names as if I was a bad person. Then when she finally settled down a bit, she invited for a coffee and to meet her sister. Unknown to me, they were hatching a plot to discredit me with our company as I was working for a Hole in the Ground Drilling outfit for weeks. I did well, t should be A-One no matter what.

First they pretended to be friends wanting to exchange recipes, then got me drunk and tried to take advantage of me so I would miss work. True, I'd been in camp for six weeks already and it might have worked, but I just drank a bit of whiskey and headed back to our camp. Their hypocritical approach had been denied as they had already drunk too much and been there quite a while, three months themselves, so it was all for the best that they received a break. After all, the contract was indeed offered to Mas'keg Mike, so have a nice sleep and pack your bags for the next flight. That is the way to take a needed break.

After a three-day break, in fairness and as compensation to one of the ladies, the position was offered, as a dishwasher or in other bushwhacker terms, the "bull cook." Yes indeed it is a full day's work and is very physically demanding.

One of the ladies claimed she could do as good as any man, well here is your chance. Some women are able to do so, while in the other camp they had a man doing all these chores. Well she was in good shape alright, so the foreman said they would give her a try. The second day there, she claimed that I fed her rotten food and tried to poison her, and then took off on the next plane. Then she wrote a letter to our drill manager asking that I be dismissed, and said she was going to complain to the health board as well.

The young ladies lies were not believed at all. Hallelujah, dear Lord.

Well we all had a laugh about that one – talk about your back stabbing...Of course that young lady was sent back to town after her long stretch in isolation. She certainly needed a break from hard work and of course had made a good honest dollar and should be thankful. Over the break they were always looking for others who like isolation.

What saved me was the fine lady geologist whom she had been roommates with. She had been on site with us from the very beginning. She was extremely good at her trade and she and I had become good friends as we had helped save the life of one of our co-workers, together that spring.

The twenty-five-year-old fellow who'd been working with us for six weeks was riding on a sleigh, heading to a site to prepare the next drill hole, with gear at his side on the start of a day. There was a chunk of the tree sticking out of the ground off by the Ski-doo snow trail and

he got stabbed by a sharp, two-inch wide, three-foot, broken tree stump, which had been previously cracked by a heavy piece of equipment. He was pulled off the sleigh by the impact, sitting with it in between his two legs and lying down in shock and screaming. His co-worker had some blankets to cover him and put pressure on the bleeding. The geologist lady quickly returned to our main camp and asked me what to do. She knew I was a Level Two first aider with experience in injury care in the workplace. I thought it over and said we need an orange marker flag, for the flight man to see the site from the air. Our location was about a mile southwest of our main area.

Next I called Buffalo Joe for a 911 ride on his next plane passing through our area.

I grabbed the first aid kit, asking with a grin if she had any tampons or Kotex that we could put to good use to stop the bleeding. Off we went up the trail a mile away, and we marked a landing for the incoming plane. The foreman arrived shortly and he marked the landing strip more efficiently for the pilot's landing. We cut the piece of tree stump slowly, left two feet of it in him, covered his injury with Kotex, and duct taped it adequately. Then we pushed his legs together as we slid him with one blanket onto the sleigh, then covered him with the other blankets. Within twenty minutes our foreman was in radio contact for the plane to land safely and he hauled the young gentleman back to Yellowknife for medical care that saved his life. We saw him in town later telling us that he got a few stitches, and he thanked us.

Well, the weeks are chipping in one at a time, with the early rise of four bells for a shift of twelve hours or more, depending on the day. There were also personal tasks of laundry for me, a little fishing now and again, and playing crib in my spare time as I enjoyed the competition.

With spring on the way, the weather was much nicer as time went by, with sunlight increasing to twenty-four hours in mid-June of the year. Of course the ice was so thick it does stick around for a little while yet.

As we were getting close to the end, on the last drill hole out at our site they had a break down. They were about five miles from the opposite end from camp, and the foreman called on the radio at two a.m. and asked if I could take a Ski-doo and go to the other company's camp to borrow 2x10-foot drill rods so they could finish off the hole. He said he would meet back at the main camp. Away I went to other camp, and pulled right up to the other foreman's tent. It is usually informal and the lights were on so I knocked and entered. Oooh doggy, there they were with one of the ladies straddled over the foreman having themselves a time. "Whoops," I said. "Yikes," they said. Hahaha then the lady closed the light with the pull string.

I apologized and told the foreman of my dilemma. "Okay," he said. "Check out the shop and take what you need." So I found the rods, loaded them up on the sleigh, and headed back to meet up with my boss. Well sir what a laugh we had in camp the next day when I told them the story.

MÂS'KÉG MIKE DABBLES IN SHARES (G)

Unlike the TV show *Cash Cab,* driving cab way up north had a slight twist. I have driven taxi primarily in the Yukon and in Ontario for a bit of time; not a whole lot. There is no difference really where you go – just take people from A to B. People who need a ride somewhere and who don't have a car might charter you also. Basically you provide a service in return for the fare. You have the right to refuse people if there is display of weapons or aggressive behaviour or bleeding that needs medical care. Call 911. It is at the driver's discretion but most times it works out okay as dispatch is there for you. You get familiar with routines, customers, and schedules. The cold winters of the north give you increased popularity. Most people pay cash for their cab fare, however, on occasion there's bartering, or postponed payment. If someone just has no money to pay fare, they may pay you a year later.

A fellow by the name of Bob Allexine and his wife, along with Hans Veros of Germany, had a big interest in City Cab. There was also good old Ed who dressed as Santa Claus every year for family groups for a good price during Christmas season. Friends of the Hanson couple I had started things on the go with advice from other

Sadly as I later drove night shift, Bob was on dialysis. During his last week we spoke about old times. City Cabs had a small fleet of about thirty-five taxis in Yellowknife back in the 1980s when I got started. There were guys from many different parts of the world that did this part-time and full-time. Many of them were miners driving for supplemental income. Twenty-four hours a day they are always looking for good drivers. I was told by a few past owners of the 1960s and '70s that though Yellowknife's population was less than 5,000, there still was a big demand for cabs back then, at forty below zero...with them being the only cab company in town.

One of the three previous owners, the German fellow Hans, who was now in his eighties, told me about disagreements amongst the owners when his wife was dispatcher and running the office. Yes, in providing public transportation she did so efficiently, in all aspects according to the law. Taxis were in big demand for all medical contracts 24/7 and the company complied with the RCMP's local bylaws about good vehicle care.

I started driving taxi back in 1981 for a gentleman who worked in the mining industry as well. Paul was from East India and he and his wife were raising a few kids way up north in Yellowknife. I was invited over to have supper of course meet the family sure learned a lot from them, raising a family and got to see their kids till they were in the workplace themselves. I drove part-time for extra income and to do something different in the evenings. My shift would started at five p.m after the day job, very often.

After my first year working at the Yellowknife Inn, driving hack on the evening shift, and saving up some cash, I got to purchase a trailer home in Trail's End trailer park, along with a taxi of my own.

Tom, the Greek fellow, was also part of the mining industry but was retiring. So with a hand shake in trust, half the cash up front, and a cold beer, he told me I could just give him the rest later that year, and he sold me my first of many taxis. I later drove for a gentleman from the former Yugoslavia and bought his taxi, which was in better shape than my first purchase. It was of a yellow color but this one was a dark-blue Chevy Impala. Where else could you make your own hours, meet interesting people, and make a decent buck?

Along with better pay, Yellowknife did have many choices available. Later on, while working in the camps using my military cooking skills, I would have a responsible driver take care of the car and then after a short break would have something to do upon my arrival back from the camp. As time went by, the two fit nicely together. There are two twelve-hour shifts available so the other driver always had a job to drive hack for a bigger dollars in the winters. The Midnight Sun made it busy later in the day...

Everything went alright for a couple of years, then one day I came back from camp and there was a rumour of new ownership and that things would change for the worse. Well I spoke to a few of the guys and there was concern. I knew the owner and he explained that indeed this was to be and there were a few people looking at buying the company. The amount needed was 300,000 dollars. Well I had three cars on the road that winter that the owner had sold to me. "Okay," he said, "you have till Monday morning." I told him of my idea about a cooperative like they had down south, so he agreed to let the drivers have first crack at it. I got together with my good friends Super D and the Maverick. We had coffee at Tim Horton's, next morning we called a lawyer we knew, and we called the drivers in one at a time, and told them of our plan. At first there was doubt as to whether we could pull it off. The coffee was coming in left, right, and center and by mid-afternoon we had sold most of the fifty shares we had discussed. Finally, the manager from Tim Horton's came to ask what was going on with all these people coming in and out. We explained things to everyone, and kept it simple and straight forward. The same rules for the road would apply, everyone would still have a job, and life would be great. Well it worked like a charm – the drivers came in and handed over $6,000 for a share, which would cover operating costs for the year and pay the lawyers' fee. The whole nine yards .So we called the owner to tell him the good news, and we celebrated with a bottle of champagne. All swell that ends well.

Only thing though was we had a big bag of checks and cash money, and who was going to hang on to it? So we did as our lawyer asked, gave back the money, and got a personal check for each share sold. In the bank line up, we were first ones in to make a deposit and later have one check made for the sale. It turned out to be split with his son but overall a done deal for that day.

INVEST IN THE NORTH'S RESOURCES (K)

When I first started to drive years ago, taxis provided courier service and pizza deliveries as well as transportation to the public all over town and at many events. They also provided transportation for medical services to people to medi-vac from the neighbouring communities, as well locally to all fire fighters. One day, I was dispatched for a trip out to the office of the Saint John's Ambulance in town. First car on the list at the taxi stand, I arrived, and a nurse asked me to pop the trunk open for a to load supplies. I opened my nice clean trunk and then we were off to the airport as she was heading to the community for an emergency and she would be back with those folks. She had been dispatched to a community to fly on Buffalo Joe's King Air Twin Otter and assist a patient for the return trip to town to be flown to the Yellowknife

Hospital. Alleluia praised to the Lord. As teamwork was in play, it worked well enough then.

Later, of course, as the population increased, EMT did improve and people were taken care of very well. There were yearly fun events that were great to attend and I was often asked to take part in some of them such as Geo Science Forum, Folk on the Rocks, Caribou Carnival, Polar Bear swim, Gumboot Rally, Raven mad day's events During these festivals the street is closed for a fun time by all, including face painting and lots of kids stuff. Once a haircut for a response to something. Stock market tips were often exchanged at the bar.

The North has folks from various parts of the world, largely because of the mining history along with good fishing and trapping being another, which is what brought Aboriginals there in the first place. Yes, having done a wee bit of skinning I helped sell them at North Bay fur trappers of Ontario as a youngster, finding out some like the polar bear furs, which come from the Great White North. Neither here nor there, per se. Bartering still exists to this day. While I drove taxi in Yellowknife people sometimes paid with carvings or furs. Fares were sometimes paid by visiting Inuit folks that did not speak English too well. I may have had one of them there as a translator, or two throat singers from the Yellowknife airport, heading to the hospital to have their newborn.

One evening at forty-below, a nice everyday couple called a taxi to go home while waiting outside at the bar at closing time. They were not dressed for the weather and the waiting time. Along with a few others, they sometimes shared transportation for a better price along, making friends on the way. Here now, I pulled over, then lightly honked the horn as the couple walked to the back door. They opened it to tell me their name, and I said yes in confirmation. While the nice lady's husband held the door open so she could hop in, a drunken fellow who was in big hurry, ran over stating that this was his taxi as he had called one too. "Check with the driver," he said and I nodded to say yes, let's go.

As they argued with the door open, and both were heading separate ways, they started having a fistfight for no reason at all. The wife asks me to call the police as her husband is getting beat up badly when all of a sudden another cab pulls up behind me. The scrapper runs over, quickly jumps in the car, and off they go. I hop out of the car and help the gentleman to sit in the back and now we are heading to Emergency. I gave them the other guy's car number to call the RCMP and have them take it from there. It's something they did take care of often... loading you in for the drunk tank ride.

TWO FOXY THIEVES (L)- 1991

It was four-thirty on an Easter morning and I had taken out a nice leg of ham to go with scalloped potatoes, peas and carrots. It was sitting outside on top of a wood box for about fifteen minutes while I prepared coffee and got the show on the road. I did hear some sort of noise but with all the rumbling of the generator and the commotion in the kitchen, I didn't pay attention. When I came out to grab the leg of ham it was gone, Holy Hanna, huh! First though, I thought one of the fellows maybe hid it on me, and I started to look for it, but no swine to be seen. The guys started to pile up in the kitchen for breakfast and no one had seen the ham. Finally, in came the foremen and announced that a couple of arctic foes were wrestling with something big in the back. Everyone started to laugh. "Do you have another?" he asked. I guessed this mystery was solved.

Curiosity got the best of me and I ran to the back and saw two white foxes chewing on leg of ham. They had rolled and pulled it about twenty feet away. They would not get a chance to finish their meal. As I approached the sneaky pair they scampered away, snarling and screeching as they disappeared in the snowy tundra. The frozen ham was trimmed and baked at 350 degree with no fear of catching rabies.

I pulled out a few steaks for a back up and was more careful in the future.

BIG MOMMA BEAR (M)- 1992

Of the many sites *MÂS'KÉG* MIKE had been at bears were always there, I was on a video in Whitehorse Yukon with a few others telling some of my for real true bear stories. Here is one!

Okay, here we go – the old Thompson Lumber mine, thirty some-odd miles east of Yellowknife, and a six-week project to reassess existing core samples and implement a follow- up drill program. The cook was a buddy of mine named Frankie, who was saving up to buy himself a truck. This time I was helping Rick, the geologist, as his assistant, breaking rock samples in half with a two-pound sledge-hammer on a vice-like contraption. Bagging and tagging samples, and stacking the core boxes in order, on the rack that were outside by the tent. I guess the prospector's course I took was finally going to pay off, hahaha. It is nice to know the type of work people do because it affects day-to-day operations. Anyways, we had been plagued by a mother and two cubs during the six weeks we were there, but no close encounter. Oh sometimes the little bears would come running out of the bush and keep going. The mother had been sighted but never came to camp, the generator noise and activity just kept them away.

This was the last day and everyone was heading out. Just the geologist and I remained waiting for Buffalo Joe with the last plane load as they would be back in a couple of hours. So no food and no weapons, just the four-wheeler ATV to be put on the plane firstly, more room. I sat on the dock while waiting for the plane. Rick was in one of the abandoned buildings with a book and a chocolate bar. Not one hour after the Twin Otter had taken off, in comes the mama bear. She must have thought we were all gone, I guess! And she must have smelled the chocolate bar the geologist was eating, because that is right where she went…up the stairs. It was then that I heard the screams. Rick tossed the chocolate bar and was making his escape out of a two-story window. Heading towards the dock yelling the bear is here as he hid behind the wooden sheds, so right away *MÂS'KÉG* MIKE fired up the four wheeler and honked the horn we had while having an eye for a good size stick along with some rocks to let her know we did not want to be her lunch.

Sure enough it was then the 600-pound female waddle out of the building so I quickly gathered some rocks to then hurriedly climb up on the ten-foot steam pipe support that meandered through camp as part of the heating system. Then I watched for my buddy and told him the bear was still in the cabin and he could make dash for it, onto a nearby roof. I told him I had some rocks and I would pin her in as he made his way to safety on a rooftop. When she came out, I yelled and threw a few rocks and she made her way into the bush, never to be seen again. Thank God as we laughed we were sure glad of that. Kept rocks till his plane came and we were on our way home again after six weeks telling our story to the pilot. Laughing away he told us he had heard many over the years. haha

CHAPTER 15
LARRY'S LUCKY SHIRT

1994-95

Now here it was 1994, and I stopped over at my friend's house to ask him how things were. Told him that I would come back for a lengthier visit than the last time, this giving us more time to relax and listen to how his family was doing, along with his wife working part of road safety off the Heart Highway of Prince George, British Columbia. Catching up over old times As we sat outside shooting the breeze, we caught up on old times.

Wasn't that just a little over ten years ago?!

Over a cold beer or two, I told my Boot Camp army buddy the story of how I got there in the first place, to visit him all the way from Yellowknife, NT in 1983, after being flown to an isolated mine site they called Salamita Mine. I had been hired by *Crawley and McCracken,* bush camp caterers that had offered me a good pay over the phone. Chef/cook position, which was surprisingly hammered down to a bull cook position for half the pay upon my arrival. I was there mostly to cover the actual cook's position as he was out on a holiday. Oh well! Did a five-week stretch with no worries. Afterwards, I spoke on the phone to the boss, and just nicely asked them to keep me in mind for the next offer, until a replacement bull cook finally made his way in. At least it was a good work out with pay check, to a future potential.

We, along with a few others, had hopped on Buffalo Joe's plane for the one-hour return flight back to the Knife and I got to sit next to a French New Brunswicker called Ben, who was an underground miner himself at this Salamita site. He had just finished a five-month stretch as a first-timer, and had gotten to experience the Great White North that year. "Holy smokes, must a made a big pay check, with the skill of yours," I said.

"Yes, with the bonus and even all the taxes."

He told me he had never been to Yellowknife before, only on the flight through it coming in, so that was where I offered him an overnight tour with a cold beer, and suggested that if he made the big bucks, he might want to live here for a bit, and that it was worth checking out. He told he enjoyed country music, and I directed him to the Gold Range Bar to grab a room upstairs, if he got lucky, and if not to come on back to grab the couch at 40 Trail's End trailer park.

Well firstly, I told him of my cute gal friend who lived in Trail's End with me. She was Boom Boom Geoffrion's niece. He could just grab the couch for a day or two, then we would look around for his own hotel or a shack. She occasionally played guitar as part time job in some of quieter places of the Yellowknife taverns, and he got to meet her.

We had our own freedom in this living quarters with respect for one another's choices at all times. Yes, with her being from Quebec City area, we had toured a few of the pubs on summer family visits.

The next day, after a bar night in the local pubs of the mining town, he was sure impressed at least with the Golden Range, a country western bar/strip joint, which made it memorable. Breakfast and coffee the next morning sitting at my trailer home at Trails End. With a new deck of cards on the table, I offered Ben the challenge of a crib game for $100 of his hard-earned money, with a cold beer chaser. Big hug and kiss to that girl friend of mine we had met heading to an early morning shift herself. Here now, Ben had the whole day to make a few calls, followed with a cold beer BBQ during twenty-four hours of sunlight at the end of the day.

Well Ben and I played a few games, drank a beer or two, and took a break with a tour around town, calling in my designated taxi buddy for the tour around. Being a slow workday we got an offer of a two-hour tour from the Greek gentleman, who smiled as he caught us up on local events. We talked about taxi business, and he stated he was a miner himself, and drove mostly to pass the time and save money for his retirement like a dozen other hackers of Yellowknife. He gave Ben some advice about the mining jobs to look for if it appealed to him later. He took us on the airport tour to show Ben the affordable campground that was popular with visitors and newcomers. Of course, he followed up with the two local mine sites; one being Giant Mines and the other being Con mines. He mentioned he had brothers out in the Stewart, BC area also looking for work.

Then off to drive around some of the lakes and river systems that hooked up to the Great Slave Lake, and we talked about a few fish stories. How every year, visitors and musicians at our festivals, along with other famous folks, fished on the Great Slave Lake for their biggest catch ever, and how most of them got it?

"Well MÂS'KÉG MIKE, you've driven hack and made a few bucks, why don't you buy a car for the road?"

"Is that an offer?" I said to him with a bit of a grin.

"Sure," he said. "Why not? Just give me five grand and you got yourself a deal right with this car."

Ben thought well, Yellowknife is where I am moving to for sure.

So a couple of hours later, upon our return, we shook hands and I owned my first City Cab #8, a high-mileage, yellow Dodge that would pass a mechanical check up at the main office on the next Monday morning. We paid our fare as we got dropped off in the downtown and we headed out for lunch at the best Chinese /Canadian restaurant. It was busy and had good daily specials. Grabbed a dinner for my girlfriend upon heading home that day.

Shooting the breeze with a few laughs, talking about what had happened, Ben asked me, "What if I get my class "F" license and partnered up with you on this cab?" Well, we discussed requirements overall; first foremost of being a good driver, and a criminal background check to get your permit, then there was an exam. On its face it was certainly doable so if it gets Okayed at the office, a partnership would be worth a try. Well Ben liked how it all sounded, so he started shopping around and planning ahead. He still needed to get his own place before month's end, and he made a few calls to line up an apartment or a hotel room in between. So with the couch available, Ben was making it all happen.

That evening, we spoke about all of this preplanned opportunity taking place.

Let's put the driving on hold, he suggested. Moving to this area he had a boat, a car, and supplies at his brother's place in Stewart, British Columbia where a mine site had gone out of business like they all do. That is how he ended up out here and he was sure glad to at least give it a try.

He was definitely moving here, so he had to go back and get his personal items in one load.

Man oh man, I told my girlfriend Chantal with a kiss and big hug what we were up to, so off we go. We both flew from YK to Prince George, BC to pick up his car, that being the closest airport, and then drove on over to Stewart. I had told him that an old army buddy of mine lived here in Prince George and asked if he minded a stop-over while in town. Sure enough, it being late in the afternoon I had a chance to make a few calls. By golly, army buddy Pvt. Stack was married with two teenage boys. He gave me his address with an overnight offer of a cold beer and talking about old times in our lives.

We ended up taking off the next day, making plans to do a return stop-over from Stewart's coastal region with at least a fish or two. Ben said he would take me out on an Alaskan king crab fishing trip I would never forget. We shared the driving all the way to the west right near the Alaska border, and were told to wait and see. Sure enough we arrived after a long, ten-hour, curved road mountain drive, with a gas stop.

We spent two good weeks with incredible fishing, and filled two orange garbage bags full of Alaskan king crab. They said I could not leave town till they took me across the US borders local pub, the Alaskan Pan Handle, to get me Hyderized.

Are you serious? Alright, so here we are at the Glacier Inn at a twenty-foot countertop, with a big bottle of Everclear and a two-ounce shot glass, with another glass filled with water. Ben told not smell or taste – it had to go down in one go and if I puked, it would cost me a round. So I did, no questions asked. Bartender flipped the glass on the counter and slid it over to light it up with his Bic lighter. Gave me a card as he told me that I just got Hyderized. So I asked him nicely with a smile to get me a full bottle then.

Along with a few striped bass from this incredible Portland Canal next door to Hyder, we had a great time visiting his two brothers and telling them about Yellowknife's mining potential for work. The two weeks went by so fast and here we were now with a boat hitched up behind Ben's Chevy Camaro.

I fished the Fraser River as we toured around the Prince George, BC community. Something about the area had an attraction that enticed me to stay. With the encouragement of their habitat conservation officials, annually, the public drops off bags of minnows of various species at various parts of the river. This lets you know their environment concerns are real. Kids are encouraged to participate that week, as well. The rugged, mountainous outdoors, along with the good fishing spots of the Chinook salmon gives many a choice for weekend outings. I had heard the rich history of the pulp and paper industry, which involved a diverse mix of people, including small groups of Aboriginals, French, English, European, and Asian population in various areas. Got to spend some time with army buddy Pvt. Stack to meet his loving wife and two full grown sons who took me golfing. Time flew by to the next generation in some of the outdoor aspects of everyday life in British Columbia. What a pleasant surprise, a chance to talk about the good days gone by. My friend said he would put me up for a while if I had a notion to look for work and stick around for a while.

Well we got an invite to a dance and it was there that I met Larry and Beth, who introduced me to a nice lady for a dance. We dated for a bit and her parents asked me to do a complete set of blood work tests before having sex. No problem, I confidently did so! While at the dance at the seniors club off the Heart Highway later that evening, Larry introduced me to a gentleman who had helped launch one of the first Tim Horton's in Canada. There sure is some history behind that. He said he'd heard I was from the North Bay area, that is one of them that I help start a few years back. We must have seen each other at the one on Cassel Street. Small world as we sure enjoyed the choices of that Tim Horton's. Small world, yes. Well, by golly sure was turning out to be fun, great place to be.

Having lost a few pounds, and now in his eighties, Larry had offered me a nice clean white shirt to wear as a sort of welcome to town gift that sure looked good on me. I wore it to the next seniors' gathering.

The gratification of doing volunteer work is great and beneficial for everyone, and it's enjoyable to do so. Sometimes the rewards are slow to come but the experience is always with you. Other times it comes back in a surprising way that pays back double or takes you places that you would least expect and allows you to meet great people while visiting places you would not normally go or be allowed. When I was offered the position as Chef at this establishment, the deal was you had to volunteer on Saturdays for the big dance. This included preparation of a light lunch and dancing with people who enjoyed an evening of socializing and dancing, after the meals. Then there was clean up time with the members of the seniors' club where everyone would pitch in to help. Same thing next week There was this couple who were a hundred years young, still dancing together. It was a pleasure to watch them dance as they received applause and many smiles from everyone.

"Well, sounds like a job offer, stop by the office to complete the paper work."

"Okay. Sounds good to me," I said.

"Can you start right now?" the lady asked.

"Yeah sure," was my reply?

I was introduced to a lady in the kitchen who was going to show me the routine. "They don't need another cook," she said to me. "They are just trying to get rid of me."

Wow didn't see that one coming.

"Yeah," she said. "They want me out of here."

Turns out she was right. Due to house politics, some of the members did not want her there.

Well, weeks and months went by and my buddy called up said New Years is indoor and outdoor Jacuzzi slug spa, bring some beers over – got a spot for you all week-end.

With a little wee bit of snow falling that evening, it was a warm enough night, and I got to meet many new folks and friends of the family.

Larry was an eighty-year-old gentleman at the time, and he was always accompanied by a nice lady named Mary Kay, AKA Katy. They had been together since his wife had passed away. Inseparable, they made a very nice couple. They were regulars at the centre in all the activities. Every day, the seniors came for lunch or activities. The centre was a fun place to work, except for the lady whom I was replacing. She had three kids to support. Sure gave me hope to live till I was a hundred years of age, though.

I drove taxi in between bush stretches, and slowly worked my way to the modern technologies of computers. Sure cleared the path for me and I kept on reading that book that encourages getting grade twelve, and it that finally happened.

Yes, well Ben did get his licence, and he made a few bucks. He kept moving on and he flew out to India to marry his wife, Cherry Blossom. Then they raised a few kids in the warmer part of Canada.

CHAPTER 16
MÂKWA KONSE

Mâs'kég Mike Meets Mâkwá Konse

Voyageur Days 2012, Mattawa, ON

My loving mother had loaned me five "grand" in 1999, to take this six-month Personal Support Worker course that is in big demand. Of course it was well needed in the Retirement City of North Bay. She knew this only too well as she had been doing it herself for over twenty years, and she coached on for do's and don'ts from the start. In applying to be a P.S.W., a full medical is, for starters. As is a recent criminal background check, as part of my seniors care training during the last month was spent in a few seniors' homes for a potential job offer. My Boot Camp army days as a graduate sure brought back fond memories of dressing clean to wear proper gear for the required occasions. My first donation of blood while in the Canadian Armed Forces was told that my blood type "A" RH negative was rare in of the 6 percent rate. Dress code came without say, gloves protective wear and your daily choice of clean attire came in three various colours; one was dark blue, the other green, and there was a wine-red. The suggested importance of doing so was to bring a spare suit as a backup just in case, along with a note pad. The training was geared to elders/seniors, and people of all age groups with disabilities of every category.- All of this training was put in place so that those that would enjoy doing so in the workplace, could do it efficiently. Learning all aspects of the trade covered the bases overall, to ensure future work for those taking this course. There were opportunities to move up in the medical field or be a Para-bus driver. That ambition was mine, as my class 4/F was in place as a safe five star driver to take people to their destination.

There was no shortage of students at this class, it was full of ladies ready to learn all aspects of care, and was a short move up to nursing if motivation fell in place. Pride with bravery as the challenge fell in place as *MÂS'KÉG MIKE was* the only man in this classroom. Upon arrival at the class we were told to find a seat for a bit of time and get to know one another, as there was a one-on-one team style for the caregiver pride as we sat in pairs at a table Being the only man in this class I was not picky, as many of the ladies were already married. So all of a sudden, with a smile, this one lady sat by one an others side. We had name cards on the table put in front of ourselves for the teachers to see. This lady and I introduced ourselves as we sat by each other asking questions about upbringing, and current places of living, as we interacted with one another throughout the day. Well, good golly, what a coincidence as we found out we had been next door neighbours in the past. She had lived in Bonfield Ontario on our farm land when she was all of five years old that is when we met. With her description of the area as we talked about our lives, it sure enough sounded familiar having met her parents what a pleasant surprise in finding out that this was all true. I knew her parents, and being the only man in a group of twenty women, it sure paid to be with someone that knew me. Her name was little D.

She was married to a loving husband with two young kids and was currently living out on a farm area and taking this course to pay the bills and add income for the support and upbringing of their children. For crying out loud we had a few good laughs as others heard that we knew each other from a while back living in the countryside on a farming town. Where many of our grandparents and family members are now getting older themselves and might be going to live at a nursing home in this northern community of Ontario. Overall, the PSW skills being in great demand did truly allow us the opportunity to be part of life in a very nice way, meeting descent folks of the medical team while being one of the branches ourselves. I was invited to meet her hubby, a French trucker of the region that I may have gone to school with at Algonquin High. They currently lived and enjoyed country style living with their children as they were passing on the raising just like we had been brought up ourselves. I looked forward in meeting her husband Makwa Kolts and the kids out in their farm yard community where I had just recently purchase fifty acres myself just a ten minute drive away. Family name to that of the man my father's sister had married meant that we were cousins as well that was just a bonus. *MÂS'KÉG MIKE* and Makwa Kolts are close to family, we have built birch bark canoe.

Well on with the class we go, we shook hands to meet one another and then all our books were put in place for a great start. We got introduced along the way to our three great teachers. Head of our class was a Registered Nurse stating she would cover all aspects for us new caregivers, from wheelchairs to lifting devices used in washing the injured, paralysed, or unconscious patients in a hospital where your skills could be put to good use. Medical terms were top of the list, in long and short form,, along with the knowledge of how to provide adequate care for an individual's needs. Nutrition, medications, and oxygen were covered with all equipment for care groups of all ages including newborn. Some of the skills were just briefly touched on, as later on your choice of facility would bring your skills to a higher level.

Being the only man in this class, a question that could be answered better by myself was handed to me directly as an advantage. This was like a dream of masculine popularity come true. One of many was dressing or undressing a patient under a blanket from clothing to a gown without actually seeing the patient totally naked, in preparation for X ray or any other medical process. This was an awesome way to spend my $5,000. A Dozen Level One and Two First Aid courses I had taken over the years sure helped me with familiarization of medical terms.

Washing a patient while still in bed was part of our training by a very skilled and courteous staff, who covered all aspects in doing so efficiently. I surely learned many things first hand to say the least. It sure felt good and there were no complaints on my end at all, in fact, I was asked to pretend a pain or other issues while getting cared for. Monitor accurately with a stethoscope use, while measuring blood pressure readings and pulse rate at a quiet time, then taking notes in abbreviation. There was training and test used on each other asking about medication, last they saw a doctor as if doing so for real on a patient. As the only man I was asked to undress to my underwear while in bed pretending to be unconscious and mourning. We often got a tour around as we visited nursing homes, local hospitals, clinics, and group homes. That was sure worth every penny, training to insure a passing grade of eighty percent.

Having put in my time, with the opportunity to take first- aid courses over the years, since my time in the military, I learned that keeping up with recent changes was a big advantage in our society today. Working in the mining industry as a chef with extra pay as a Level Two First-aider from courses I'D taken before this, had helped me be part of saving lives while in isolation. You name it, everything from bear bites to fractures, a finger amputation or deep lacerations of the body. These were some very worthy components to all of this new skill that I looked forward to achieving, and soon using in the public as a PSW. to be part of the medical team.

We learned to work under pressure in many aspects of Health Care, following the rules and keeping the door open after the six months, with a chance to use these skills in the workforce. In caring for others properly it was important to do so for yourself daily, being on time, eating properly, control of your weight, coping with stress levels. Indeed, the motto of being a good listener was part of the communication process with one; namely without interruption, criticizing, or offering advice. W proudly followed all the rules and regulations in dedicated team-style fashion. Matching our strides daily as a team- taking notes, and yes identifying and accepting differences to avoid conflict as we build a strong foundation in our workplace. Should there be any issues, they would they would be resolved by the upper ranks. My mother loaned me her van to take my stepfather Cookie, an eighty-eight year-old war Veteran on weekend tours. Actually using the skills from the training, helped me build my confidence of high hopes for my future,

On my fifth month, with an 80% grade and doing very well. I had nearly gone through most of the course in a smooth way with a few minor discrepancies, what with being the only man there. Being a born and raised Frenchman and learning the English version of medical terms in a short bit of time had in itself been difficult. But overall, the course was a marvellous, educational, awesome, and worthy achievement. I felt proud of my five star driving skill background as a Para bus driver opportunity or Ambulance drive, as that was yet another course later on. I had received offers from one organization in Yellowknife with the Association for Community Living as a Driver/ Facilitator working with the disabled, my God. Both my loving grandmothers of mine were at a local nursing homes, one of them where I put in a total of three weeks training. It sure looked good as well when I got to meet a lovely aunt of ours that was ninety-nine years of age, hallelujah. I was feeling good about all I had learned for my future in passing this course, and was now on the fifth month, with less than four weeks to go, for my certificate of achievement the best ever for my new career.

All of a sudden, on a mid-week morning, this smirk-faced lady teacher/ counsellor, not a local RN. Approached me with an emotional attitude. Having just arrived to the workplace parking lot on a early Monday morning at the start of our day. She asked me if I was the one who had tampered with her car as the hood of it had lifted up on her while she was driving over here on the Trans-Canada highway a few miles away. Less than an hour before! Honestly, not being the only one that did not particularly care for this wild lady, I never would have even thought of doing something as evil as that. She accused me in front of a few other students rather than office discussion with the board of directors, which is how we do it in civilized society, in Canada. I said something about not even having transportation for myself in the first place or at this exact time living with family. As a five star driver I cared for the safety of others while on the road a large part of my life working in public transportation.

That was it, I quit right on the spot to just walk away from this evil lady's accusation, knowingly that our Lord worked in mysterious ways. With my five months accreditation and 80% passing marks nicely given to me at this College's main office, I once again stated that it was surely not what they had taught me. Native Education Training College accepted my hours in confirmation to complete the exam with another months' time at two local Nursing Homes to finally drive a Para- Bus at nearby senior's home on a part time basis to completion.

Through local library, by word of mouth we put our efforts together for our lineage search to validity our DNA for our Algonquin Treaty card. Meeting other families we had found out as time went by in our genealogy that we were cousins right from the start of our whole twelve generations. With a few years difference of age her husband, Makwa Kolts, told me that his second name, is of the Algonquin word meaning bear. With a smile, he being a distant cousin himself as my father's sister had married one of his dear uncles. We had kept our relations amongst each other till now. I told him of my bushman's name since the 1980's being *MÂS'KÉG MIKE* which only meant Swamp Man. Well he started to laugh as did his wife and kids.

"Where in tarnation did you get that name?"

I have used it in many places, as there were at least five others with the same first and last name this was given to me out at an isolated bush camp near Pickel Lake Ontario.

With the many of us in both Canada and the US that have the same first and last, this nickname tactic minimizes the identifying process. Having worked in the isolated regions of Canada for many years, and often hired over the phone or radio, they knew who they were talking to.

As we sat around the table he was telling me about our lineage and how due to the Algonquin Treaty that was still falling into -place, he was accepted as a Métis in the provinces of Quebec and Nova Scotia. DNA with lineage proof played a big role today, as did the use of our family names being part of the treaty settlement just over 240 years ago. Ontario's regulations were that you had to be of pure First Nation blood for the last five generations or you lost your right to be called Métis or even an Aboriginal. Agreements of each Treaty from the start we that all the Chiefs, Wives, and Braves with their children would be cared for as they were a large part of the population that had been encouraged to mix with other ethnicities had majorly kept Canada in growing. Just to say that 240 years beforehand, the forked tongue of many had made political promises that were questionable. Mixed blood was worth a try, and was at least better than killing one another over the large space of Canada. This,- was just too obvious to many. Métis Louis

Riel had hit his prime time being a non-discriminant, spiritual, and political leader ahead of his time.

Thanks to our Algonquin Chief, of course Anahareo's friends and family grandson Mr. Lalonde for permission to display Makwa Kolts and MÂS'KÉG MIKE's hand built birch bark canoe by the Mattawa Museum, then later at the Capital Theatre of North Bay, ON. Dedication to both writers Gertrude Bernard "Anahareo" and Archie Stansfeld "Grey Owl" booklet of life history and genealogy. Attending the 100th anniversary of her birth in 2006.

Anahareo's memorable 2006 gathering of her 100th anniversary of her birth in Mattawa, Ontario as Makawa Kolt and *MÂS'KÉG MIKE* attended with many others. We had our paddle signed at the display of our hand built teepee and birch bark canoe with ride offers for the kids. Attendance of many Aboriginal Elders one well known William Commanda. of Maniwaki QB. Playing her great role in the Grey Owl movie, actress Annie Gallipeau was also there.

Anyhow, marriages, births, and deaths, along with other registrations of most Canadians were kept in order by churches, hospitals, funeral homes, librarian's, and newspapers to name a few. Most importantly, overall the direct lineage from the day you were born to that of your Algonquin First Nation relations, whether it be five generations to twelve the Treaty start of 240 beforehand, covered that space from now till 2014 to its settlement. One of the best proofs of documentation was the marriage certificates at your local church, or communicating with others who had some already in place. Pleasantly enough, many of us were related to others who moved from one small community to the next over time. Online information was helpful in finding as much as you could at one time. Upon our discussion to find our direct lineage,, it became one of our projects and we communicated with the North Bay Librarian Mr. Boisvert, who showed us the path through the many books to read. Resources from, Rene Jette's genealogy families of Quebec, L'Abbe C. Tanguay of Canadian Families, Dictionary of Pioneers, Quebec's National archives. Immigrants who came to Canada, Pioneers of US/ Captives, French and Native North American Marriages 1600- 1800. To name a few that cover all of ours. The Catholic foundations kept track reasonably well, although sometimes churches burnt or were rebuilt, when we asked for our family name on both sides, we'd then search sometimes through second or third marriages in a book listing of alphabetical order. We searched church or community locations, reviewing different regions in order to see who the grandparents were and on and on it went. Catholics keeping lots of records. Well after putting some together with a few left to go. I told my friend that Yellowknife had lots to offer this way as there were Ouellettes with the same lineage as mine, so perhaps they would help me in being part of the Métis group and would give me some idea. Twelve generations of research was a big project for newbies. Saying goodbye to pay off my loans, I had met an Algonquin Elder while in Yellowknife. He would help me find out more about all this.

CHAPTER 17

TWO BECOME ONE

August 2006 – North Bay, Ontario

With the hope of perpetuating my ancestors' bloodline, I decided to go on a search to find unfold the mystery of who was destined to be my other half. Since I wasn't active on the dating arena I thought checking online will be a good start. One of Google's suggestions was Cherry Blossoms and there I was signing up for a one-month membership. 30 days came and nothing exciting happened. Just as I was ready to give up and discontinue my membership, an email found its way into my mailbox. After I read what she said in her profile, this brown-skinned damsel strummed my heartstrings and played the most beautiful music I ever heard.

"I'm going to the Philippines," I informed my parents.

They were genuinely surprised. "Are you crazy? You've only seen this person through the computer and you think you're in love? You've never even met her!" They exclaimed.

"And that's the reason why I'm going over there to meet her and her family."

"Well, if that's what you've decided, just make sure to stay safe."

13 months of daily emails, long conversations on the phone, animated handwritten letters and greeting cards for every special occasion, and text messages were the main ingredients of this cybernetic love affair.

Here are a few at Christmas

It was good to hear your voice on Christmas Day.

Hi sweetheart

First, I wish you a very merry Christmas!
It's Christmas Day today there, right?

Second, I'm glad you got your permit. That's
good news indeed. Congratulatons.

All my love

Hi Iris angel baby

Thanks for your understanding and patience with
help of God our faith, relationship is growing
stronger all the time I will be thinking of you
every day call you on Sunday 9 am my time but
if I get another job up north I may not be able
to call right away. Don't worry I will get a hold
of you be happy life is good God is great.

Love you Michael

September 2007 – Manila, Philippines

I advised ma chérie that I will fly to her beautiful country, the Philippines, to meet her and her family. It was something I really looked forward to but I couldn't deny I had some reservations. Suddenly I was barraged with doubtful thoughts and questions that could only be answered once I am in front of her. I honestly believed all that she said but that nagging uncertainty of "what ifs" kept creeping back into my thoughts.

"Once you see her, make sure she is a genuine lady and not a man pretending to be a lady," said my Asian seatmate on the plane on the last leg of my flight to Manila. Well, so much of hopes and believing. I was more determined than ever to meet up with her and get down to business, ha-ha! "I heard a few sad and twisted stories of men in similar situations as you so be very careful," he added.

"Ladies and gentlemen, welcome to Ninoy Aquino International Airport in Manila. Local time is four o'clock in the morning. For your safety and comfort please remain seated until the seatbelt sign is turned off..." Finally, the longest flight I was ever on has come to an end, I thought. 14 hours, man, that was brutal! I badly needed a nice warm shower and a good sleep to make me think straight.

I stood up from where I was seated and my cramped legs felt relieved. As soon as I stepped off the plane into the jet bridge, the rush of hot, humid air hit me in the face. Whoa! I then knew what I was getting myself into in this part of the world. If you are a frequent international flyer, you're familiar with the process of going through Customs and Immigration. Since this was my first, it was one good reason to travel light. No line-ups in the carousel waiting for your luggage and hassle-free transfers. I realized that my flight arrived an hour earlier than expected. Ma chérie was surprised when she got a phone call from me as they were just walking from the parking area toward the arrival area. I approached a Filipino guy and asked if I could use his phone and I handed him $20 Canadian dollars later. Ma chérie tells me later that she would only believe that I was real once she'd see me in person. So here I was standing by the coffee shop as I said and there she stood looking candid and lovely just like in the photographs she sent me. Then a there was a brief awkward moment as we kind of said hello unsure of what to do next. So I swept her into my arms and gave her a lip lock. She introduced me to her sister, Madonna, and Boyet, her brother-in-law, who were part of the welcoming party.

The ride from the airport to our hotel suite in Makati City took about 15 minutes and I was ready to hit the shower. "I did not come all the way from Canada just to shake her hand," I said to Madonna and Boyet. They gave me knowing looks and a sly smile then left us to ourselves. From the time I held ma chérie in my arms I knew she wasn't a fake and I never let her go since then.

I woke up to an orange glow in the window. It was sunset and I realized I slept the day away. As we left our hotel, we were greeted by a young guy who tried to sell me knickknacks, cigarettes, condoms and what not. That's different, I thought. We walked down the street and came upon a lighted food cart. Dinner consisted of hotdogs and barbecued pork, a common street food in Manila, and probably the rest of the country. Ma chérie asked if I wanted to try the "balut" "It is duck egg with a developing chick inside", she told me. I chickened out after she told me what it was. "I believe I tasted that before in the Yukon during a visit to one of my Filipino friends. But I'm not a fan of it though." I stated.

Ma chérie and I went back to our hotel suite after popping the last morsel of supper into my mouth and went through our to-do list for the following day. We were ready to tackle the day after a hearty breakfast. Before we left the hotel, I psyched myself up to battle the heat outside.

First stop was the Canadian embassy located in Ayala Avenue. Ma chérie was trying her best to get us a taxi but we got refused each time. Why? Many of the taxi drivers would charge us a flat rate instead of using their meter. Ma chérie knew the embassy was only several blocks away but because the heat would be too much for me, a taxi was the most convenient mode of transportation. But be ready to haggle or you may end up paying more. I was glad that ma chérie was with me. She argued that the rates they cab drivers were asking were just too much just because she was with a white guy. Being in foreign soil, I let her do the talking and negotiating. Finally a kind-heart cabbie stopped for us and agreed that we pay by the meter. He sure got a good tip.

The Canadian embassy was brimming with people even before the doors opened. All I really needed there was a document certifying my legal capacity to marry in the Philippines before we could apply for a marriage license. While we were waiting for my name to be called, a gentleman seated at the corner was speaking with his fiancée with a heavy accent. I couldn't help myself so I approached him and in a friendly voice asked him "Sir, if I may ask, whereabouts are you from?" "I'm originally from Greece but now a Canadian citizen. I'm here to marry this lady," he said gesturing to a young lady who smiled shyly at me. "This is my third marriage and I am 80 years old. Thank goodness for Viagra!" He bellowed. "Right on!" I replied.

The wait wasn't that long. I heard my name called and we walked to the window where I was asked a few verifying questions. Then the officer handed me my certificate officially stamped and signed.

Boyet and Peter, Madonna's boyfriend, picked us up at the embassy and they took us to the Philippine's largest and one of the world's 10 largest malls, the SM Mall of Asia. I had a feast of the eyes but shopping was never my cup of tea. I was more into tasting Filipino delicacies so off we went to the food court area and I tried almost everything that looked weird, unusual and whatever tickled my fancy. What a blast!

After two days of touring the country's capital, we flew to Cagayan de Oro, one of the major cities of Mindanao. From the airport, we decided to spend the night as it was getting late and I preferred to travel during the day for security reasons. And because I traveled light, I was running low on clean clothes. I asked where the nearest Laundromat was. Luckily there was one at the end of the shopping mall conveniently interconnected with our hotel. Unluckily, we picked up our laundry and my new pair of jeans were switched with an old pair. It was a long day and I was ready for bed. Our room was clean and the bed was comfortable. I picked up a calling card on our way up so I could call my folks back home in Canada. For sure Dad and the rest would be wondering how I was doing. We hired a cab for the two and a half-hour ride to ma chérie's hometown in the province of Bukidnon.

Bukidnon My (New) Home

Ma chérie's family are such good people and I am so lucky to be a part of them. They gave me a warm welcome and everybody was excited to meet me. Me being the first foreigner married into this family, I felt honoured. One Filipino way of making a guest feel at home is to share a bottle or two of Tanduay using a single glass passed around until he bottle is empty or until the party's over. Tanduay is, as they claim, a national icon among the distilled spirits market. A staple during special occasions.

I did not waste time. I faced ma chérie's parents and formally asked to marry their daughter. Then I came down on my knee and presented the ring to my chérie in front of her parents and popped the question in Tagalog which I rehearsed for months. *"Pakasalan mo ba ako, mahal ko?"* She thought it was corny but said yes, of course.

For the remainder of my stay in Bukidnon. We booked a tiny one-bedroom cottage at the Edlimar Spring Resort which was a five-minute walk from ma chérie's parents' home. Edlimar, as it is commonly called, is the first spring resort in Bukidnon owned and managed by a very charming couple, Mr. Ed Marquez and his wife, Lilia. They are very good family friends of ma chérie's family and what better way to show our appreciation for their outstanding hospitality and friendship than to request them to stand as our wedding sponsors?

I married my lovely wife in the Philippines in 2007. That's a bamboo!

The Wedding

My wife and I were married twice – once with the judge and the second one in the Catholic Church. The plan was to be married twice in just one day. Why twice? There was no certainty that we would get a date booked at the church as the finalization of the plans was only a couple of weeks before. So we made sure we booked with the judge just in case. We had to fit all these excitement within the three weeks of my vacation. It was indeed a whirlwind romance for the wedding to happen in such short notice.

The invitations were limited to immediate family and ma chérie's closest friends. I had no representation from my side of the family so I made sure someone was taking a video of the church wedding. This video sure was brought back for my friends and family to see afterwards.

Ma chérie looked stunning in her wedding gown. I looked dashing in my suit. We overcame the jitters and had a beautiful ceremony. The reception was held at a hall big enough to accommodate us in the same resort we were staying at. The *"lechon"* or the roasted pig made a delectable centrepiece. I can honestly say that everybody had a good time singing away with the karaoke machine and drinking the fruit punch I prepared myself. There was beer *grande* for the big boys, of course. After all the guests have gone, our bellies full and our hearts singing, I carried my new bride in my arms inside our cottage for an intimate wedding night.

The civil wedding with the judge took place the day before the church wedding. It was scheduled for one o'clock in the afternoon. There was a brief interview with the judge before the wedding. At the time I thought it was a violation to my privacy. The lady judge questioned my intentions of marrying a Filipino lady. It drove me to the point that I wanted to about the wedding. Too much? Well, I felt insulted with the tone of her questioning. Later, I realized the judge was just doing her job. However, the ceremony at the church the next day made me forget this unusual encounter.

Canada – Her New Home Sweet Home

With my wife at the Joint Task Force North headquarters in Yellowknife, NT where I worked as a Commissionaire.

My vacation was over as quickly as it began. I was back to work with the Commissionaires and back to my usual routine. I then started the process of sponsoring my wife to come to Canada with the help of some friends. While waiting, I kept myself busy by driving taxi part-time in the evenings and at one point I enrolled in business course at the Academy of Learning.

Eight months later in August 2008 my wife landed in Canadian soil. I told my wife that I will meet her at the Edmonton International Airport wearing her the cowboy hat given to me by her father. Two very good friends of mine will accompany me. My friend, Joe, came up with a funny idea that he'd like to wear the cowboy hat instead. Here comes ma chérie walking down the stairs her eyes scouring the crowd for the cowboy hat. "I see the hat but how come it's worn by somebody I don't know?" she wondered. Behind my friend, I tried to hide while we were slowly approaching her. The when she got close enough, I came out from behind Joe, wrapped her in my arms and I planted a big kiss on her lips. We all had a big laugh.

The four of us went out for dinner after which Joe and his wife, Barb, drove us to the Nisku Inn where I booked a honeymoon suite. As we entered our room, we were greeted with a bottle of champagne sitting in a bucket of ice cubes and a dozen red roses for ma chérie. We relaxed in the Jacuzzi while sipping champagne. The following day we flew to North Bay so my family could meet ma chérie. As promised, I took ma chérie to Niagara Falls for a honeymoon getaway and had lots of memorable experiences to last us a lifetime.

We visited family and friends together that very memorable time in our life.

Praise be to our Lord

CHAPTER 18
COOKIE

Here now, MÂS'KÉG had returned home in 1999 for a time to reflect on taking Personal Support Worker schooling, and of course to be around family as my parents were getting on in years. Sure nice to visit my fun relatives and be in familiar surroundings.

Having spoken with a gentleman only over the phone a few times, I was told he was on his way to being a hundred years of age. Now here I was to finally get to meet Ernest and hear of the many stories he had to tell me. With my PC skills in place, it sure made life easier for me, with emails and unlimited online info to do research, which is how we live these days. Fact finding was a rewarding task that would leave living legacy as part of history's fateful turns. Cookies was a nick-name given to Ernest Cook, as his family name was Crookedneck.

I received my personal support worker certificate in 2006

So here we are at the breakfast table at a private section of my sister's abode, Ernest now telling me the story of his life and showing me some of his documents, as I briefly told him about the writing my life story. So here is some information to include me in there as well.

I knew my mother had been a caregiver for a large part of her life and had been taking care of this eighty-eight-year-old war veteran named Ernest Cook, aka Cookie, along with his loving wife Elvira for a good eight year period. Sadly when Cookie's wife passed away he was grief stricken by the loss of his loving partner of fifty years. Good God, she had passed on before him. Cookie had now ended up in the hospital for more than a week, with major depression issues over his loving wife's death. P.T.S.D was one of his major difficulties before his loving wife's final days. Throughout visits during these sad moments, and later when family expressed their condolences at her funeral, he still attended church during his recovery time. Now, going on his second week he just wanted to return home and continue with his life as a widower, as he knew it would not be the same for him anymore. Of course my mother had been for visits all along.

Rules and regulations tell the story of how he got here now going on his first year, living with my mother. Due to the legality process and largely because of his condition, he could only be released to a next of kin, yes a family member. However Cookie wanted just to regain his independence and deal with his grief in the comfort of his own home. This was at the seniors' apartment he had shared with his spouse and of course my mother the caregiver for now a few good many years. While my mother was visiting him at the hospital during his recovery he would tell her he just wanted to get the hell out of this hospital but they would not let him go.

No one else seemed to care as he was up there in age, and being told that he may just be transferred to a local nursing home in a short bit of time. According to hospital policies, the time was close for him to be signed out, into the care of a family member that was it. With not many family members to choose from North Bay, where he still sure preferred to stay, he had been well taken care of by this nice lady till now. Since my mother was the closest one to him, he was sure glad to see her every day upon her visits.

A few of his local Legion friends had come to see him as well during this difficult time of his loving wife had passed away. He spoke to a few of his old army buddies while they came over for the visits about the outcome of our governmental system that was all part of our life. You like your caregiver, know she is trustworthy. she makes sure you take your medication in everyday life, and knows you better than anyone else. She knows how to handle your Post Dramatic Stress Disorder better than anyone. Well just ask your caregiver to marry you, before they stick you in a nursing home against your will by gosh.

Made sense, by golly, to keep receiving adequate care that was the best idea ever. So that is when it all took place. Ernest's mind was still sharp in making choices for himself, one of them was being able to keep going to the local Legion to play crib, and be around some of his long-time war veteran buddies, and enjoy life. Taken to his loving wife's funeral, he saw everything taken care of at the church and the visits afterwards with the cards and condolences of the many people that knew them over the years, and that calmed him some. He was telling them of the plan he was choosing and he honestly felt it was all for the best.

At the end of the last week at the hospital, largely due to his medications, not just anyone could easily care for him. So it was then that Cookie called my mom and he asked her to marry him, and care for him like she had done for them in the past because there was no way he was going to a nursing home. At eighty-eight years of age, Ernie still had a very clear understanding of what was going to take place and happen to him if he did not do something soon. My mother thought it over and discussed it with family, and she felt okay in doing so. In front of friends and witnesses, off to the altar they both went. Cookie came to be cared for and now live with our family, at a house close to the church so he could live with some partial independence, to reasonably enjoy transportation and health care from home. So here I am as he continues telling you my story is that good enough for your book. On a Sunday drive he had gotten to meet my own father who lived nearby with a nice lady.

Now living at this house, Ernest got to spend time with our family, seeing our young nephews grow living and being part of all the birthdays, Christmas, and Thanksgiving till the New Year, for a total of six years that he truly enjoyed. A semi-private unit of the house gave him privacy, for moments when he thought of his past, listening to country music or of an old cassette tape of him playing the violin singing at events in Sudbury, over forty years earlier.

His circulation was getting bad, a massage at the start of his day with a little Dr. Scholl's massage device sure helped. My mom made sure he had his three healthy meals along with regular visits to doctors as needed providing him with a visit to the nearby Legion.

Of his daily medications, one was from where he had suffered a mild heart attack with a bit of angina issues, and he was on also a diabetic's diet, taking seven other pills daily? This medication did keep him reasonably healthy. He had to eat a low fat diet, with small portions, and now and again he liked to cheat at a stopover for a scoop of ice cream. He was easy going for a walk on his good days. Having served his country in WWII and now a veteran who still enjoyed going to our local Legion, he attended military gatherings in tribute to our Canadian soldiers who had served our country and now continue to do so. Suffering from the torments of P.T.S.D., some of his days were spent right in bed or quietly looking at his albums for hours in memory of the good days of his fifty years of marriage, playing music, and the hunting that he had done for many years. Deaf on one side more than the other adjustment and battery needed to insure he could actually hear you not just smiling or nodding his head in agreement.

Despite his Post Traumatic Stress Disorder, he now and again enjoyed telling us some stories of those forgotten days, while with other vets sometimes, as I went with him to our Branch #23 Legion of North Bay. A few folks called him Cookies then one day he was telling me how this name had come into play, largely due to his original family name this was liked. Upon my visit I got to spend a lot of time with Cookie in giving my mother a break to go play her favourite game of BINGO, she sure was very lucky as others would whisper as she walked by. There is a lucky bingo player of this region she has won the jack pot many times so try and sit close to her it might bring you some luck. That is why she always came in at the last minute.

Knowing I had spent time in the military she would ask me to stay over with Ernest for such a day to see and experience his behaviour to Post Dramatic Stress, and watch him. Privacy of a two bedroom unit with a few looking out for him at all times he really liked living where he was at in this particular part of his life telling us our Lord looked out for him. With a Personal Support Worker course about to take place my mother stated, son you still do not have your grade twelve you should take this six months course to insure work demand in the future. With time and visit to family course falling in place with a loan coming my way in trust I did apply to do so with encouragement wright through. Firsthand experience helping Ernest with driving my mothers vehicle, of course I walked to stay healthy was sure glad to have met cousins in class to spend time with local families in the research of my Aboriginal lineage.

This day Cookie's early morning start all in his mind with a blank look, now as if he was in WWII sitting behind a wooden stump at the ship to shoreline defense. Time stood still as if in another part of the world, or was he just out of his mind as this was all happening again right that very moment while he was sitting in bed, and I was sitting on a chair watching out so he did not fall over. It was explained to me over time that part of the coastal defense was on the shore line of Juno Beach – during the Dieppe raid, France expected an attack and the soldiers patiently waited in readiness with prayer to our Lord. Talking to me as if we were both right there on that day.

Having been at Camp -X this top secret area of Ontario during WW II reminders in his stories. He would tell me, if he would run out of bullets or had them in a dangerous area, he'd use a hand signal to get someone to toss him some, while they were being shot at. Soldiers he had just met, layed dead close by in that moment. With a tear in his eye saying he was sure glad to be here today. After a rest he showed me his albums, along with telling me more military of his stories of WWII. While in Ontario's base sections he had trained as a sharp shooter in defense of his country he would show off as if holding a riffle in his hands. With respect to all others who were training, he had really enjoyed serving his country. Ernest loved to wear his military uniform, proudly wearing the medals during ceremony events, which I got to participate with him on a few occasions. Sitting by his side was an honest moment of pride for myself.

Cookie as Pte. Coucroche RHLT-B.38181 (Ontario)

Shared with others as they told there stories a little later while sitting at a table in the Legion as if one of there army buddies getting ready to just make it through the day. You were in the military yourself, eh! Yes I told him, *MÂS'KÉG* MIKE was my nickname ha ha !

On my time off, having taken Ernest Cook to visit a few longtime friends at their homes or at the one of the many Legions in Ontario, he was a lucky man most of the time. Stopping at a store he'd like to buy raffle or scratch tickets, and there were pretty good odds he would win a few bucks. When I took him to one of his favourite sports, the hockey games, he would smile and tell me he was feeling lucky. "Go put my name on the door prizes would you!" he would ask me, so I did. Luck is a mysterious part of life in my opinion as I got to experience it from my mother now from Cookie telling me directly of feeling such a moment.

Cookie as right fielder 2 of his baseball team, 1940.

Sure enough, as the first period started, they called his name. Since he'd taken off his hearing device, I told him that they had called his name, so I went to pick up the North Bay Battalion hockey team shirt he'd just won and he smiled with a high five, holding the shirt as he gave it to me. "Hold your mind to roll with lucky thoughts in everyday life," he said.

Alright Cookie, I will try.

Another afternoon my mom and good old Cookie sat at a luncheon table having a coffee and chatting with the café owner, a lovely lady who I knew from our local Indian Friendship Center. Looking at me with a big smile, she told us the basket was nearly out of tickets but that that there might be a good one still in there. "Here is twenty dollars," said Cookie. "Grab what's left to bring at the table for something to do while having coffee.

As they opened the tickets my mother shows a win of $100.00, saying she's buying us all the lunch. Cookie still had a few left to slowly look for a match, when all of a sudden he pipes up and says there's no way my mother was paying for lunch because he had just won $1000.

Even the owner came over to have a look at the winning ticket from the last little bit that had been left in the barrel. We all laughed in getting ready for a bannock luncheon special that day

Halfway through my Personal Support Worker training, on a long weekend in the year 2000 my mother needed a little break herself. I lived by myself in a rental unit a five-minute walk from her house. With a military background myself this was hands on experience in care of a war vet that sure was worth my efforts. The life-story bonus was all the more fun. With a van, all his medical apparatus, and the new *MÂS'KÉG MIKE's* safe driving skills, we were off to visit some long-time friends of his out in the Sudbury mining area. This was where he had been brought up and he had lived there for half of his life, working in the lumber mills with many years in the mining industry himself just before his WWII time. What a great experience to take Cookie to visit other Legions of the area with a few war veteran friends of his from some mining sites, who told us the stories of how he'd gotten there. He had received an invitation from the Falcon Bridge mining owners, as he was one of their oldest employees that was still alive. With my cell phone, a note pad, and an address book of locations, on the road we go.

Both of us liked Aboriginal family history and he was now proudly telling me of his upbringing on the Ojibway Reserve at Whitefish Lake, just around the corner of Naughton, Ontario. He said that his mother was originally of the Beaucage Reserve part of the North Bay Nipissing area. Ernest's family name was Coucroche aka Crooked-neck, initially baptized with the name Commada. My stepfather's mom's maiden name was "Commanda," that name came from a French Canadian's term for a commander, a well-known Algonquin Chief of the Nipissing region who had led his tribe in the Canadian war settlement of 1812. Transportation was by canoes, dogs, and sleighs with horses when allowed. Farming was still part of raising a family for those with that name in that time period. Hunting and fur trapping on either side of the river's forty mile expanse was now part of marriage, and of course raising kids thereafter. He had brought his family from the Quebec side of the Ottawa River to the Nipissing regions of Ontario.

Grandfather of many a well-known Algonquin named MKISHINAATIK, Chief Francis Commandant (Rotten Wood) had occupied the area with hunting and trapping back from the 1830s. Of his many grandchildren, one by the name of Chief Semo Commanda had a community named after him, along with a nearby lake just off rural routes of Ontario's Hwy 522. Online, yes by word of mouth, but mostly with the accurate registries of family history at the local libraries, I was encouraged by Ernest to do more research for the book. Of course his family had lived all along Nipissing, sharing the large mass of land for a time. Listed on the 1871 Ontario Census as a "Voyageur," Louis Shawanakesi Commandant, married to Mani Anne Kijikawokwe Johnson, had a daughter named Marie Catrin Commandant. Born 1892 on the Ottawa River.

As Cookie and I exchanged stories on our drives out in the year 2000, he now told me that he had left home one summer's day when he was about thirteen, with a handful of peanuts in his pockets and no shoes on his feet. His gear was packed on a dried piece of wood called a 'bindlestiff" with his red and white kerchief on a stick, which was his travel kit to tote about as he made his way down the road of the unknown to find a new path in life. Once again faith did smile upon Ernest as he stood outside, near a lumber mill right by a feeding shack in a small community called Benny, Ontario, looking for a bite to eat, in about 1918, or so he stated. The owner Mr. Cecile's wife had taken Ernest in as a kitchen helper to meet the wood hauling crew of twenty, which hauled by horse, for a good size mill that was in big demand for that time. Within one year his next step was to be trained as part of the mill's crew, bundling pickets all the way on through this busy mill operation. Cookie closely became part of the large Cecil family like a nephew, an uncle, or a brother to all the younger kids, which is how they still refer to him today with kindness and respect. With one of our stop-overs of this old mill site area no longer in operation, a few folks still live there who actually knew him from those days. As we pulled up for directions at a local convenience store, sure enough we met one of his mill friends.

Over the years, Cookie had built a fishing camp out on Onaping Lake between Sault Ste. Marie, ON and Sudbury as he loved fishing in isolated spots. It was a one-hour boat ride from the shore line to this island and the camp he built a little bit at a time with his loving wife of fifty years. He suffered Post Traumatic Stress Disorder even now while on his daily medication for a few issues one of them being Alzheimer's. He had adjustable hearing devices in both ears. As he often sat in the passenger seat, with a smile I would ask him now and then if he could here alright, especially when he was startled by a loud noise from a secondary source. I saw him shut them off on one side or both. He would point to a sign or a building by a lake area to let me know that he had fished or hunted on that side of the mill a few years before. He'd ridden on the top of a huge winter sleigh pulled by two strong horses, which he pointed out to me in one of his pictures. They were pulling a large load of logs, and heading for the mill.

"That's me on the top of the pile there."Reliving his life on a summer drive around to where he use to go with others sure made these moments of his great storytelling as if we were right there. He would ask me when my next time off for our next drive out was. That same day, we headed out to downtown Sudbury near the Royal Canadian Legion Lockerby Branch # 564 and we went in for a little washroom break to talk to a few members about some of his close family war buddies, as he called them. Speaking to a nice lady at the Legion branch, we bought a coffee with a few tickets and he asked her what about my brother Weldon Cecile or my in-law Clifford Montgomery. Well, she smiled stating that one of his WWII veteran relatives was at a seniors' home just up the road, and she gave us the address.

Thank you much Ma'am, we are off and away to visit a few folks of my relations – have a wonderful day.

Sure enough with the address given to us, we arrived at this nursing home in the Big Nickel and we got buzzed in to the main door, mentioned who we were looking for, and showed our ID card. Have a seat at the entrance way, and she called one of the caregivers for an update, to then hand us a little pin-on visitor name tag as we were taken to Clifford's room in this very swanky place. The vets met at the door, yes, they had not seen each other in a long time. We got offered a beverage at a seating area at a table. This place had sleeping quarters with a small kitchen and washroom, and they ate at the main entrance with a wonderful view of the outdoors in this down-town area. As they spoke, I just sat listening to them talking together and drinking some water. This being a long weekend, I was asked if I would drive all the way to Chapleau, ON where Ernest had a home built hunting camp, right by a lake in the middle of nowhere.- This is an isolated moose hunting area and you're sure to get one every year. Well here we are getting close to lunch so we got offered a meal as the vets continued their discussion of the days back when. They moved around some, and I made sure Ernest took all of his medication at meal time.

We ended our meal, getting ready to say good-bye on this Saturday long weekend, and shaking hands with a big smile, Ernest asked me if I was up to driving Highway 101 to take him all the way to Chapleau, ON for supper, to visit one of the Cecile-family nephews of his. he had not seen for many years. Well hold on, I'll call Mom so we can both speak to her to let her know what in tar-nation we are doing today, heading for a five hours drive, way farther up north than I had expected to.

Sure enough, with a couple of calls we had permission to do so from everyone. That is when Cookies asked me, are you up for the drive to Chapleau ? We have a place to stay overnight for a nice rest with a good home cooked meal. Alright then, off we go with good byes and handshakes trying to get there before dark on this road with a lot of wild game; a big time hunting location, looking to visit another close family member he had not seen for a few years. Hells bells we finally make it there in one piece with a few washroom breaks in between. We look in the phone book for the address to drive to as a surprise visit, and knock on the door.

First let's go around the corner on this little gravel road while it is still light to see his hunting camp,-sleeping quarters, just off of the bridge on Lisgar Street. Sure enough it is still there with the key hiding in the same place, to just go inside for a look around. It was right by a little lake that is part of a river system hunting spot. We had another little bush style washroom visit to go back to this here Chapleau town. On our way back I had a little stop on the side of the road by a sign that read Brunswick First Nation Reserve near a gas station for us to fuel up at. I saw a fellow with a moose in the back of his truck coming in by the pumps, so I grabbed my camera, Ernest asked me where I was going.

I am going to shoot a moose," I said, then pulled out the camera from behind my back.

"Are you serious?" he asked, as he walked by the truck that had the moose.

"Yes!" I said.-"With a moose call and my camera, hahaha!"

As he walks by me the owner grabs my camera saying he would take our picture together.-Right on thank you, here is one for the book eh!

Gassed up we come knocking on the door of a son of the Cecile's from the mill days. We had a nice visit with a good supper, and they offered us a basement suite to go for a sleep.

Letting the others be to talk about old times, I went for a good rest to later wake up as Ernest told me we had a good place to sleep at a local motel for an early departure tomorrow back to Sudbury if I was okay with that, with a sleep over at Mr. Cecile's niece that he now had an address for us to visit. All good, no problem. Sure enough, the following day nothing but nice weather coming our way, as we skipped town on back to the Big Nickel belt of Ontario. We were expected after our five-hour drive at the home of this lovely lady who lived in a fourth floor apartment. We drove right by the Whitefish Lake Reserve that he was raised on himself.

"Well want to do a little stop, over while we're close by, to see where I grew up?"

"Sure, sounds good to me. Let us pull in there for a tour."

Cookie knew my capability to safely be on the road having driven 3000 miles to Yellowknife in four to five days more than once. He would ask if I felt alright and I'd say not to worry, I was enjoying this. Driving taxi of 7 days a week, between bush runs this was easy.

"Just trying to make sure you get some rest and move around as we stop for breaks. "I said. "We have the carry on washroom in the back of the van should you need it." As we stopped over at the Whitefish Lake Reserve for a quick visit I asked a pedestrian where the Chief's office was.

We stopped over to get permission to do a little tour. We had brought his military documents with us, along with a marriage registry to show the family a little of the research that I had done at the North Bay Library.I promised him that I would make a dedication to Ernest and that he would be in a chapter of my book as -WWII veteran, step-father Ernest Cook /Coucroche/ Commanda. He like many others was of both the Algonquin and Ojibway lineage as they grew up next to one another. Cathrine / Katline Commanda, born in 1880's passes away in 1913 and lived on # 10 Nippissing First Nation of Ontario, the Beaucage Reserve and was one of the many children of Algonquin parents. Ernest's mother Cathline had married an Ojibway man of the nearby Whitefish Lake Reserve, by the name of Ozawance Coucroche – aka Crookedneck that had lived off the LaVase River Portage system that was historically a meeting place for many during the

Voyageur Days canoe route just before the 1850's settlement, it has a stone monument marking it in place that was built by the Landry's a cousin the of mine in the 1980's.One of our drives was at this southern area of Ontario that Cookie had apparently travelled by canoe all the way from the Whitefish Lake Reserve with his dad, Angus Coucroche. We took about a two and half hour - drive to Parry Sound, on this weekend, so he could show me part of the canoe routes he had taken as a young fellow from Whitefish Lake Reserve, through on to Killarney Park, all the way to the Lake Michigan shore line. It used to take a little over a week to get there. A vehicle would have been a lot easier for sure. It was early fall as we drove into Parry Sound to an open house parking lot of a corn roast Cookie instructed me to pull up right by the serving area, and there stood a tall gentleman.

Cookie rolled the window down and said to the guy."You sure look like Bobby Orr"

The guy replied, "I use to play hockey with him when I was a kid I am his younger brother"

We all had a good laugh.

Cookie told me that as a boy he spent time on Parry Island of Ontario as he escorted an aging Angus to see an eye specialist to try and restore his sight. Sadly there was nothing that could be done for him, Ernest got to meet new people out there as we drove by one of the only swivel bridges in Ontario where he had spent two seasons and was familiar with the area. He told me that a swivel bridge to and from the island was not a common sight, and that he had caught a big trout just below it. He'd needed help to pull it in, as he tooted his horn.

Cookie had lived in many people's houses and hearts over his ninety-nine years of life, as he did in our family for a brief time of about six years. He was still alert to the last day. Was visited by many friends at his

My mother our family got to hold the flag with many were visiting him when he passed away, with two separate gatherings at Mcguinty's Funeral Home of North Bay ON. Ladies Auxiliary of our Legion Branch #23 were notified of his passing away, as details were put in place for a traditional ceremony of paying respect to a WW II veteran Ernest Cook. Full house that day as each line of tribute was done in fairness to all attending. Military dress code of proudly wearing all there medals, traditionally honoured and lead by the Comrade in Charge. Group lead in placing poppy's on him with a flag, then followed by all other Comrades at the casket to pay homage, say good bye to Cookie. As part of the ceremony the Sargeant-At -Arms walked over to salute as he placed his poppy to then his position at the end of the casket. Silence in the act of Remembrance followed with prayer as they paid respect to us who stood and cried. Invitation at the local legion just a short walk away on that cold day in November to spend time together in thanks after this awesome ceremony at Cookie's funeral.

I wrote my life story with a dedication to my stepfather Ernest Cook / Coucroche dit Commanda, who had joined the Canadian Armed Forces in Sudbury at thirty-five years of age- He was being a sharp shooting rifleman. He was sent to Wentworth Regiment to be part of the Royal Hamilton Light Infantry (RHLI) in 1939.

Pvt. Ernest Coucroche was transferred R.C.A.S.C., member to see action landing at Brest in June 1940, as part of the 1st Canadian Brigade to assist France against the German Blitz-Kreig. The Royal Hamilton Light Infantry Heritage Museum has a yearly display Battle of the Dieppe France.

He being a 1942 survivor, he was later sent to Borden, ON with a star defense medal of Italy, France, and Germany.

The Rilleys, "Always Ready" battalion - Honourably discharged, after a thumb amputation Oct. 02, 1945.

I officially registered him online with the "Aboriginal Veterans Tribute honour List" and geneanet.org.

The New Year of 2000 brought Ernest aka Cookie to the North Bay Library where we found the registry of his first wedding. Registry by Hubert Houle, Northern District branch of ON- 1984 /Volume 6 publication # 59: Ernest Courocha had married Lillian Turner, GS JE 08 - 07 1947 in Beaucage, ON-witness to this event was the following who later became an adoptive parent by the name of Angus did proved accordingly.

Wedding names, Angus Coucroche and Louisa Commanda - Alexander Turner and Sara Rivet, Field, ON. With many other book searches, as we got encouraged and found more reading for me to check. This was now making its way to the year 2004. Cookie thanked me for spending time in research on his behalf. The originality of his family name sure was one good laugh as he told me his first wife did not like the name too much.

By well-known author William A. Read, one of his book's family names.

"Louisiana Place Names of Indian Origin"

Of a French settlement in Lower Louisiana, in the US, the Indian Tribe that cultivated this well liked vegetable called the potato pumpkin is what my stepfather's family name became. From US growers of a horn-shaped pumpkin type called in French "ginaumont," English "cushaw." Later changed to "Coucroche" aka "crooked-neck" "Forty years later upon marriage his name was shortened to "Cook"

Of a well-known Canadian Folk Song – called maringouin aka "The mosquito." If the mosquito wakes you with its song, or tickles your ear with its stinger. Well Voyageur to learn that this is a devil's sign that he sings around your body to get your soul.

Wampum belts with stories were passed on, and we learned about family names' importance in keeping track till now.

Copyrights of 1990 the Algonquin Regiment Association (New edition second printing 2003)

Algonquin WWII war vets, my family are listed in a recent book called "Warpath" 1939 -1945.

Published under direction Algonquin Regiment Veteran Association, Major C.L. Cassidy, D.S.O.

COOKIE was not in this book However a few of my Ouellette uncles were listed, in the Nominal Roll of the First Battalion, Algonquin Regiment 1940- 46. North Bay Ontario Canada.

My stepfather Cookie. 2004

We received two e-mail invitation from the lady Chief at the Whitefish Lake Reserve of Ontario near Naughton. Brought with us pictures with all of the above information for a memorable luncheon gathering. This following Elder gathering invitation was filled with direct info of Cookie's past and he just enjoyed himself.

"COOKIE" an ***Aboriginal Veteran***

May 25 2004 (email gathering offer)

Chief G.S. would like to know if you and Ernest would be interested in coming to a luncheon here at Whitefish Lake Reserve Arrangements can be made with the other elder's to set up a date and time at your convenience. Please let us know as soon as possible, so that arrangements can be made with the elders of Whitefish Lake. On Wednesdays here in the First Nation, we hold luncheons and the majority of community seniors attend.

If interested and have a date in mind,
please feel free to e-mail me back.

The following Wednesday, Cookie and MÂS'KÉG MIKE both attended this memorable luncheon with a lovely lady, an elder by the name of Nora King of nearly the same age, with whom Cookie had been neighbours with during his early years. It was as if it was just yesterday growing up near one another in the early 1910's, side by side and each speaking their mother tongue of the Ojibway language, about their families of the area at the Whitefish Lake

Reserve of Ontario. She was part of a book herself, that described many of those marvellous moments in writing and she told us where we could find it in the library. They sat together having coffee and talking about those days, and she showed us pictures where they had a cabin nearby nearby ninety years before.

We found out Cookie's birth date was five years earlier than he thought, he was born in 1905. Using registries along with verbal information about marriages in confirmation with valid ID and pictures we learned that Ernest's mother truly was the daughter of a Commnada and had an uncle just off the Beaucage Reserve of North Bay Ontario. Uncle Gabriel Commanda who had been WWI vet with Grey Owl, the husband of Anahareo at the time living around the Temagami Ontario doing a little trapping being part of the mining industry for the area. He had joined the military as well got married in North Bay ON just 60 miles south of Temagami. I have the military certificate to prove the research being done in truly accurate.

Cookie at the Whitefish Lake Reserve with Nora King. 2004

Nora gave us directions that day to Cookie's mother's burial site. Ernest recalled his teaching with prayer and a feather in hand of his Ojibway upbringing.. He told me of his given name, "Gayaashk wag" the "Seagull". We stood by her burial site with a song in respect to communicate with our Creator for guidance and wisdom to having brought us here.

In two of our visits we were loaned an informative, 1982 book to read called, ***"Whitefish Lake Ojibway Memories, No 06"*** by writer Edwin Higgins. Available at the North Bay, ON library, as well. AMICUS - Canadian, online. Further research in regards to Cookie's family history was also given to us at this great gathering. *MÂS'KÉG MIKE* started putting facts all together. This book was based on the Treaty settlement of 1850,s with accurate info.

Book pages 76 and 77, Nora who is in the book, the Elder lady to who we spoke to Nora King tells us where on the La Vase river they lived, then on page 195 and 196, Appendix VII have the "List of Band Members" Family members with 2 of the children on the proper order "List B 1891" Treaty settlement name of past band members on page 196. Family name Coucroche is also part of the nearby Mississauga Reserve #8, Joseph perhaps the brother of Ozawance. John Ball married to Rose Coucroche 1917, church St. Famille Blind River Algoma, ON.

Omiscosance, Angus (widow) Band No. 42 D.O.B. O05/09/1877 (OAS)

Father's Name, __ Ozawance, Appendix VII list of Band Members 1850, to 1891, and 1979. List "A "1850 Ozawance Coucroche with three kids. Pay list of Robinson Treaty Annuities Whitefish Lake Band # 6 past band members, List "B": Listed in 1891 September 01, 1891 Coucroche, Angus/Paishiguin- Coucroche/ Papakina Isabelle. Commandant name is with Angus' sister as well, this lady named Isabelle Coucroche was married to Francois Xavier Commandant. Their son Abraham Commandant who was born 31-08-1903, then baptized in Naughton, ON. This lady Aunt Isabelle married to Kathline's brother, Francis Commandant. Cookie's older half-brother Angus Coucroche and wife Catherine Masinigijig where at this nephew's baptism in 1903, in the Precious Blood Cathedral of Sault Ste. Marie ON. Angus later changed or misspelled his name in 1936 at the Cathedral of Precious Blood, likely a misspelling due to lack of education. Cookie told me he could count well enough however, he did have a drinking problem where he was slowly losing his sight in middle age. As Cookie's father had passed away. Angus his half-brother now became his stepfather. The family name of Masinigijig is also in the appendix of the settlement. Spelt Mazenejkijik -Mahzenekezhik.

Registered Indian Record –

Found and shown to us at their office, this info then sent to me by email

This is the document we received at our gathering on that day, of course with ID proof in confirmation In return we gave the Chief copies of all his military postings along with pictures.

May 18, 2004 e-mail

RE: Ernest Coucroche

Census

Angus Omiscosance band number
was # 42 or can be # 40.

1919 Census - Angus was 46, Ernest 13

1914 Census - wife died February 1913

1906 Census Ernest born June 26, 1905

1901 Census Both Angus and
wife are Roman Catholic

Self-31, wife (21 to 65 inclusive) not sure of the date of birth approximated this shows just an approximate age, for perhaps people at that time were not there as they lived on the lands. Should you be trapping or living on either side of the Ottawa River border of ON and QC.

Registered Indian Record

Omiscosance, Angus (widow), Band No. 42

D.O.B. 05/09/1877

(OAS) Father's Name – Ozawance - Coucroche,

Census

Census never says wife's name. According to Chief's E-mail, her name was, Katline Commanda. Angus and sister in-law Louisa Commanda are both listed at Ernest's wedding in in the Sudbury district, she being of the Beuacage Ontario Reserve is also on the census.

No Commanda's on record, nor is her age revealed. No sister or other child is reflected in the records. All documents state that Ernest was illegal. Of this we are without any information.

To look for family history, you might want to: Yes we did contact all the phone numbers.

Write to the Diocese of Sault Ste. Marie, as Census records state that both were Roman Catholics. The Diocese keeps relatively good records of all baptisms.

Contact the Department of Indian and Northern Affairs Canada at http://www.inac.gc.ca, and request Treaty Documents Prior to 1901-1850 (1850 is when the Robinson Huron Treaty was signed). We do have copies of these, but are hard to read as photocopying was primitive.

National Archives of Canada or National Library of Canada http://www.collectionscanada.ca/02/020202_e.html, this has a link for Aboriginals to research history.

Ask Ernest where his mother was originally from, then contact that particular First Nation to look in their Census Records (Lands and Trust Officer could research).AAND 1 800 567 9604. A lovable person, with the good fortune of having a lot of friends and extended family through marriages.

Head of Households and Strays- 1871 Census of Canada, ancestry.com

Child of Louis Shawankesi Commandant and Mani (Mari) Ann Kijikawokwe Dit Johnson: listed Canada, Mari Catrin Commandant Household Female Birth Year 1880 Birthplace QC/ON Religion Catholic Source Information: Census Place Baskatong and Lytton and Sicotte, Unorganized Territory, Ottawa, Quebec Family History Library Film 1375861 NA Film Number C-13225 District 97 Sub-district TT Division 4 Page Number 9 Household Number 62

Saving the best for last, as Ernest Cook/ Coucroche Commada was part of our family, a WWII veteran who lived to be near a hundred years of age, sure is worth the mention. My thanks and a dedication to the Ojibway/Algonquins and his cousin Elder William Commanda. We also had verbal confirmation of his upbringing from Nora King of the same age group who told us the by word of mouth facts and I am able to put it in writing in their behalf with permission as we visited the reserve for a memory lane tour as we drove around. Later that day at his mother's burial site he taught me a prayer. And with a feather in his hand he told me of his Ojibway upbringing and his given name "Gayaashk wag." the "Seagull". As we stood by her burial site with a song in respect communicating with our Creator for guidance and wisdom. Prayer to our Lord as we lay on the lap of creation in the spirit of our body by a sacred fire in gathering with praise at the circle of life, to one another in preserving the earth.

Cookie with MÂS'KÉG Mike

What a great afternoon as we drove around, stopping over in areas by the lakeside where he had lived as a child. Combined with the stories in our discussion we had seen pictures of the morning gathering of that day. There was mention of his mother being Commanda of the Beaucage Reserve of Ontario, which was not so far away close to the La Vase River pathway.

In our discussion of my book in progress, Ernest said he liked how I did my research and told me he had a few pictures of his army days. Then he nicely asked me to include him in my book if I did not mind or had room, well. Surprising to me, his research was done directly with people of his age who offered to tell us directly, and supported it with paperwork they gave to us. My father Percival asked me how it all went, as he had not attended, nicely showed him a few pictures to let him know he was being mentioned in the publication of my book. My father was very impressed as he had met this veteran himself, always smiled at one another with a hand shake.

With Elder William Commanda in his home at Maniwaki, QC. 2010

Cookie passed away in November 2004, at the North Bay. Hospital with a great worthy military style funeral shortly after. The military ceremony by his casket was attended by family and close friends with a Legion gathering afterwards as he had done for others buried near his first loving wife of fifty years. Ernest Cook had given me a beret all of his many certificates of achievement for their use along the way to be part of a chapter. Born in 1905 in Bocage, Ontario and named Ernest Commanda, then adopted by the Coucroche family misspelled as Coucrocha. In the first marriage registry at the Library dated August 08, 1947 an impressive family name was changed to Cook in Oct. 20, 1950. It was done at the Supreme Court District of Sudbury, Ontario and then was published in the local Gazette by the solicitor.

Upon my wife's and my visit in 2010 to Maniwaki, QC at Elder William Commanda's fun-loving home by invite, we walked around outside the gathering area of his yard enjoying the view. Elder William, with a lovely lady friend, proudly showed my wife and me the area for the annual gatherings for the culture of peace that were part of his yard. Afterwards, having a coffee with bit of a snack he showed us his wampum belt of oral tradition, along with his many written awards of today. One of them was the "Gold Key" he had proudly received from Ottawa, which acknowledges the First Nations' paths alongside of one another in the healing circle for our Mother Earth's protection, as it has been for many generations. Showed him my treaty card and he filled in a few of the blanks for us, as we asked questions about Ernest. Here now 2010, MÂS'KÉG MIKE with his Honey of a wife having done a little research for the book had an invitation to visit well known Elder William Commada of Maniwaki Quebec. Being of the near same age as Cookie who had now past away, showing him pictures they looked like brothers he laughed we shook hands as he stated that we were cousins. He explained to us that Cookies mother and her sister where nanny's, care giver in Bocage area they would then meet a man to have children as the French River was a means of travel in them there days with a many birch bark canoes of course bigger boats like the "Chief Commanda". Elder William told me, I know only to well as I built a few myself similar to the one you and Makwa Kolts did at the gathering in Mattawa. Highway 17 railway did cost money was still new with lots of gravel road that made it difficult to travel without a car back then. That afternoon we thank them for this great visit.

Meegwetch Thank you

THE FINAL INSPECTION The soldier stood and faced God, Which must always come to pass He hoped his shoes were shining, Just as brightly as his brass. "Step forward now, you soldier, How shall I deal with you? Have you always turned the other cheek? To My Church have you been true?" The soldier squared his shoulders and said: "No, Lord, I guess I ain't. Because those of us who carry guns, Can't always be a saint." "I've had to work most Sundays, And at times my talk was tough. And sometimes I've been violent, Because the world is awfully rough. But, I never took a penny, That wasn't mine to keep. Though I worked a lot of overtime, When the bills got just too steep. And I never passed a cry for help, Though at times I shook with fear. And sometimes, God, forgive me, I've wept unmanly tears. I know I don't deserve a place, Among the people here. They never wanted me around, Except to calm their fears. If you've a place for me here, Lord, It needn't be so grand. I never expected or had too much, But if you don't, I'll understand. There was a silence all around the throne, Where the saints had often trod. As the soldier waited quietly, For the judgement of his God.

Step forward now, you soldier, You've borne your burdens well. Walk peacefully on Heaven's streets, You've done your time in Hell." Author Unknown~

Cookie RHLI Wentworth Regiment Always Ready

Regiment WWII Algonquin

In case I missed anything...

***Adventures of** MÂS'KÉG MIKE*

During cabin Fever in my neck of the woods, I became a Bushwhacker Artisan now put my PC skills to good use thanks to FriesenPress!

````A farming lad with an audacious spirit, how he got there is a mystery even to him. *MÂS'KÉG MIKE* had braved the wilds with a knife in one hand and an axe in the other; he chewed up Terra for sheer entertainment. Weather was not much different than back home, or so he thought! Not until he first experienced the many cold ones in Yellowknife, NT Winter of 1979 did he concur that there could be no work place colder than this, yet in the North it was more isolated. The barren lands seemed cruel, yet it was full of wildlife, and the unscathed summer beauty of Canada, eh! Yes for a good paycheck, like a few others before and many after in the heat of desperation, *MÂS'KÉG* did what there was to do, the best he could. Army-trained Chef-cook by trade, he was often put on the automatic bear watch duty where nobody ever went hungry. Algonquin High School had given him the skill to wield a hammer, and to build a boat or shelter. Percival my loving dad had work gloves and safety toe for his boys. Uncle Viny had given me the Toronto 401 driving skill, to hack on the black ice at forty below.

If you're not from that far-up Land of the Midnight Sun you have no idea about life in the remote areas, surely around the "Knife", where *MÂS'KÉG* spent half of his life. I lived in the"Old Town" area Yellowknife, my shack was just around the corner of Ragged Ass RD. Best ever lived at a six room party place in Yellowknife called the "DOG HOUSE". Gained acceptance by the Metis of Yellowknife to be part of Aboriginal events, I experienced cultural Dene lifestyle of the many generations first handed, having dated a few local Dene ladies to one who at least told me we had a child together and not to worry at all. I used my meat cutter skill of a sharp knife being part of the annual harvest, skinning wild game from a nearby trap-line by a few, to then smoke store prepare the meats as they have done for many generations.
````

Yellowknife Campus promoted my wealth of knowledge as a Camp Cooking Instructor to our local Dene, later a memorable course to Inuit student's way up in Kugluktuk Nunavut. In thanks I did take the Academy of Learning class of the Entrepreneurial Business Course.

There are Stories and tall truthful tales of life experienced with northern Aboriginals, Canadians and those of the mining trade from various world parts all talking to one another about where the next mine is or discussion of drunken bar issues.

Sometime hired at a local pub for a work in an isolated area, with one fearless hour to get ready to fly in with Buffalo Joe, with my boss's word of, "You'll see when you get there" to perhaps learn new skills. Labourers, either in town or out, are still greatly in big demand as required in isolated northern mining region of Canada. Diamonds rush days are still in place, a formidable chance to learn many hands on trades in the use of equipment, dirt, grease, gas and oil chased with a soap bar. A mighty small percentage of the population lives in the Northern region, but I made a few good friends on the long summer days in the Land of the Midnight Sun. Sadly, more than a dozen have passed away either of ill health, injury, or of suicide. It was Big time worth a visit back home to see family now and again, to soothe my spirit. All my bros and of course my very pretty sister did make it there for a few years. Others visit in the summer, but sadly many do leave when the temperatures hits forty below zero. Should you be seen by the RCMP wobbling on the streets especially at 40 below, they would just nicely take you to the drunk tank for an overnight stay with coffee in the morning, yes I was there once.

A fearless few do stay and shape their lives to spend enough time as one of true Northerners themselves, or raised their kids there for the generations of this day. One big time enjoyment for many was to have a cold one, then dance at the Strange Range bar all night. Yes, there were three major bars that had famous singers from many parts come in the summer. There were great summer festivals, winter carnivals, and monthly events for all age groups.

Sometimes while speaking to someone in regards to work opportunities, get the courage to try something new or different, as this may become your line of work in the future. Only if you're up to it risk-wise, there are no guarantees per say however, openings are always available due to the lack of individuals who are brave enough to give it a shot. There's air travel to the isolated mining region workplaces.

I travelled here and there for over twenty- five years on the Precambrian Shield to make bread in more ways than one. There are true stories about searching for and finding gold, Silver, diamonds, oil and gas. I fed the crews with fresh fish and backyard hunting, in between, while working with some buds and budettes from coast to coast here and there. As a Level Two first aider, I patched up a few bear bites, along with workplace injuries for the adventurous workers of my day as a needed process of mining and exploration. Bottom line, driving hack sure helped me pay for my worthy education along allowing me at meeting many well-known people between bush runs, like, Actress Margo Kidder, Musicians, Politicians, sport gals and men like, David Suzuki, and Pope John Paul II as he kissed the Airport tarmac of Yellowknife in 1984. I worked with Buffalo Joe and used his services a 911 Ice Pilot, worked with well-known local Cowboy Joe of Yellowknife. Lived a few years in the Yukon, where *MÂS'KÉG owned* a cabin near the Yukon River of the Tagish community just south of Whitehorse where I did pan for gold, filmed to tell my bear stories at the annual festival. Got to ride and make friends with a few mushers met the first American lady that won the race. Saskatchewan Metis chum and *MÂS'KÉG did* a two week Top of the World Highway drive. And I got to hear. Stompin' Tom Conner's music three times here and there.

As a registered member of the Mattawa North Bay Algonquin First Nation, participant memorable Powwow gathering of 2006. I got to meet Elder William Commanda at Anahareo's (100th birthday celebration-) and Grey Owls, both well-known Canadian authors. Part of that awesome moment was the yearly Voyageur Day event *MÂS'KÉG* and Makwa Kolts displayed a homemade birch bark canoe that we had just build from scratch. We took pictures at our teepee meeting, and family and friends signed our paddle, right by the sixteen-foot pine statue of legendary Big Joe Mufferaw, with an awesome view of the Mattawa /Ottawa River systems. Like a few others, Grey Owl's grandson encouraged me finish to write my book.

I have the picture with the paddle in hand to prove it, so here it is for you to read.

The Spirit of Truth in prayer firmly helped me to complete of MÂS'KÉG MIKE'*s Adventures.*

The anticipation of the book publication that is about to take place gives me hope.

Wikipedia's, Definition of Treaties:

> Treaties can be loosely compared to contracts;
> both are means of willing parties assuming
> obligations among themselves, and a party to
> either that fails to live up to their obligations

can be held liable under international law. The contract that is still in existence from past tense.

A treaty is an official, expressly written agreement that states use to legally bind themselves. A treaty is the official document. Which expresses that agreement in words; and it is also the objective outcome of a ceremonial occasion which acknowledges the parties and their defined relationships.

Not long after, in heading back to the Great White North of Yellowknife, I still had my class-four taxi-man, chauffeur's permit that always came across to pay off my bills and lined me up with bush work. Did a couple of stretches in between. Got to meet new folks who lived in a community of the mining industry, being like a miniature version of Toronto. The North Slave Métis Alliance had accepted me as Métis allowing me to that lived life in a fun filled way to say the least. Wouldn't you know it, I spent time asking around and did get some idea in my research. Then sure enough, back to work in the bush camp where I met a Metis line cutter of St-Pierre Ville, Manitoba who told me that his aunt was a genealogical specialist with the Historical Society of Saint Boniface, Manitoba. No kidding, give me your family names on both sides, she will do it for you for a cost. As I gave him my grandparents' name, one being a St. Pierre, he said he was related as he was a St. Pierre himself. Wow, what a coincidence.

By use of tact and firmness in overcoming obstacles I made continued progress with help from my wife, family, church members and March of Dimes meticulously achieving ambitions at the start of each day, to the next. Take notes, do not miss taking pills, eat a healthy diet with exercise. As a proud member of the Legion in support of our Troops I have permission from our Chief and Captain of the Algonquin Regiment of North Bay to wear my beret and feathers at a powwow gathering.

As part of a registered Section
35 signed by the Queen

Research of many books at libraries online of course word of mouth sure tells you the truth.

The Canadian Constitution (1982) recognizes the existence of three main native groups in Canada: First Nations, Inuit, and Métis. Since the 1970s, Canadians have used the expression First Nations to designate peoples previously known as "North American Indians," that is, any native peoples who are neither Inuit nor Métis. Today, Canada's native peoples speak some fifty different languages, most of which are spoken nowhere else in the world. Bill C-21

Canada's population being about 11 million. One million served in the military in (1939 -1945)

Population of Ontario, Canada 1941 was 3,787, 655 now 13.5 Million 38.4 %

The story of the Algonquin Regiment 1939 -1945 *WARPATH* G.L. Cassidy

Stepfather's family name was in the following books from the North Bay, ON. Public Library.

Written by Author, Edwin Higgins was "1982 Whitfish Lake Indian Reserve, N0. 6"

Atikameksheng Anishnawbeck First Nation. Author, Wayne F. Lebel of West Nipissing Ouest

Family and faith joined peoples by rivers and lakes to all of our Canadian and US society of today. Canadian census list how many people were here to first divide our current borders. Population of half a million in the"War of 1812" is a large part of what comprise Canada today.

Honour to all Veterans Past and Present for our Freedom

The Universal Declaration of Human Rights' (UDHR) declaration of United Nations General Assembly of December 10, 1948 in Paris, states that after WWII it is a global expression for all human beings. According to our census, the population of Canada with Aboriginals of the 1700th to 1800th had increased to nearly a hundred thousand, to say that safety in numbers sure did apply. The scattering of the Red Man played a large role in supporting the freedom of life as many of their wives would now be a mixture of European descent to become Metis. Aboriginal ladies who married or had children with a white man, lost thei rights without say. All the ladies played a large role in making it happen to where we stand today. While borders between Canada and the USA had now slowly fallen into place, the same folks were part of that finalization for their own benefits as Canadians and their future families had fought over the same values just before all of this. Particularly with the Seven Years War of 1812 for final Canada/US boarder Canada outcome. All of the rigmarole of scattered paper work for the said political agreements that had made it happen, hurled people from various parts of the world for a fresh start to some free land. Here now where the 1850's population had increased by over two million with the open door to be free, with property available to all newcomers. Simultaneously, finally putting into place a few of the Aboriginal reserve agreements to accommodate fur trade, lumber, harvest, hunting, and mining, namely for the future of all the families of those who had fought. Despite the Algonquin's' treaty that had not yet been signed for various reasons back then, there were no major highways, aviation, or railway systems in place from A to B. It was still mostly horse trails and waterways. The existing Algonquin Park of today in itself is part of the reason it was not quickly signed for a small acreage at that time. Here is the story of my family with the historical facts of many other Canadians for the pride of our existence till now. A true story, which is still being told from long ago. Crown land sales should have been held off till all treaties were fairly settled first.

Europeans attempted to impose their own language on many indigenous peoples of Canada and the U S. Language of course being the major means of communicating, our history suggests the differing of tongues often twisted many of its words to a new definition or a translation that was often misunderstood. Of the many languages known to have existed, many are now extinct and this leaves but a few speakers of the one's left. Although sign language is not a part of any of our existing languages of today, we all use it every day. Having five years under my belt as a certified P.S.W. - I learned that many of us already use sign language. All over the world it has been handed down to all ages. Treaties had a valid purpose for all over time as a lack of education gave them no choice but to follow the path of a signature on an agreement. Respected Aboriginal Chiefs with well-known names uses a translator who was then later exchanged through a marriage. Bilingualism, along with writing skills for some came with education passed on to them by word of mouth. Some signed their names with an image of their clan sometimes being, "Bear" - "Wolf" – "Bird" or just a number. In their own language they often just used names that were suggested, or signed with an X. Included were adopted names.

Family names for First Nations of Canada and the U.S. were properly used for ID purposes, being part of the written agreements between these nations in the binding of ties established in treaties. Like the Europeans' transition of a family name when brought to Canada and the U.S. for a very good reasons as many still have the same first and last names. Liked or disliked by Aboriginals dialect, names was often being a given by choice as a title of the trade as a family name or of their location just likes streets getting one of their names. The Ouellette name is used to state that they live by clear water area eg dit auclaire in French. Leaders became identified so to be referenced in the papers for the many purposes of registry. Lack of education or misspelling often brought humour in its historical use, to sometimes be asked where your name comes from. Genealogy and DNA proves to be all from a small group.

A Little drop of Rain by Junie Prunie – Nov. 2005

A little drop of rain
No bigger than a tear
Can make a puddle wide
If it rains here and near
A tiny puddle here
That sits in a little dip
Was made for shiny boots
With a fancy rubber grip
Many little drips
Raining down and near
Can make a splashy sound
Just listen and you'll hear
A pair of shiny boots
To splash through puddles wide
Are calling out to you
You'll miss the fun inside
Let's stomp each puddle through
Come out, come out and play
The sun has gone away
It's a rainy sort of day

An Ode to All Cats

There is a cat called Bjarne
Who figures he's a one cat army
But in truth he's just a smarmy
No good rotten little feline
Who when the chips are down
Would make a cheetah's beeline
To his mistress's fluffy gown
And then make like he owned the town
Then there's a Ginger Seville
Who acts like he could kill?
Cause he likes to feel the thrill
Of stalking that Bjarn cat
Across the living room floor
Then growls like a banshee bat
And cause Bjarn cat to soar
In the air and ask for more, more....
But a cat is a cat for all of that
You can love them, and even pat
Them and they might have even sat
Upon your lap all soft and purry
If you know not in their mind
You will look like contagious scurvy
As they ponder thoughts of their kind
By wandering off showing their fluffy behind

Michael Mann 28th November, 2001

About the Author

Michael Ouellette spent a large part of his life traveling throughout North America, eventually traveling half way around the world to marry his lifetime love in the Philippines. Independent at an early age, jobs took him to places he came to love. He had worked many safety hours in isolated mining regions of Canada eagerly feeding crews with his cooking skills which he learned while in the military, combined with his Level II first aid skills to help save workers' lives. A prospector's license earned him a few shares in the mining industry. Catering to the public was like second nature to him. In between bush runs, he operated and owned a few taxis or driving the para-bus for seniors and the handicapped. He also worked for the Corp of Commissionaires stationed at the Joint Task Force North (JTFN) Headquarters in Yellowknife. Traumatic brain injury might have briefly thrown him off-course, but he remains a force that will not be held down. The stories in this book portrays his zest for life. Furthermore, he hopes this book sheds light on how safety of workers especially in isolated mine sites be a critical onus of employers that should not be taken lightly.

Printed in Canada